Michael Imperato

# An Introduction to Z

Chartwell-Bratt      Studentlitteratur

**British Library Cataloguing-in-Publication Data**
A catalogue record for this book is available from the British Library.

All rights reserved. No part of this publication may be reproduced or
transmitted in any form or by any means, electronic or mechanical,
including photocopying, recording, or any information storage and retrieval system,
without permission in writing from the publisher.

© Michael Imperato and Chartwell-Bratt 1991.

Chartwell-Bratt (Publishing and Training) Ltd
ISBN 0-86238-289-0

Printed in Sweden
Studentlitteratur, Lund
ISBN 91-44-35111-9

| Printing | 1 2 3 4 5 6 7 8 9 10 | 1996 95 94 93 92 91 |
|---|---|---|

# Contents

## 1 INTRODUCTION 1

Enter "Software Engineering" .................................................. 2
The Communication Problem ...................................................... 3

### 1.1 Formal Specification Languages 4

What are Formal Specifications? ............................................... 4
The Need for Formal Methods in Software Engineering ......... 5
The Problems With Formal Specifications ................................ 6
Approaches to Formal Specifications ....................................... 7

### 1.2 Introducing Z 8

The Mathematical Foundations of Z ......................................... 9
Why Bother Learning Z? ............................................................ 9
The Disadvantages of Z ............................................................ 10
The Future of Z ......................................................................... 11

### 1.3 Further Reading 11

# 2 SET THEORY     13

    An Introduction to Objects ............................................................. 14
    Notation ........................................................................................... 15
    Set Display ...................................................................................... 15

## 2.1 Fundamental Concepts in Set Theory     17

    Membership .................................................................................... 17
    Equality ........................................................................................... 18
    Cardinality ...................................................................................... 19
    The Empty Set ............................................................................... 20
    Set Inclusion .................................................................................. 21
    Subsets ........................................................................................... 22
    Finite Sets ...................................................................................... 23
    Special Sets .................................................................................... 23
    Set Comprehension ....................................................................... 24
    Examples ........................................................................................ 27

## 2.2 Set Manipulation     28

    Set Union ....................................................................................... 28
    Set Intersection ............................................................................. 29
    Set Difference ................................................................................ 29
    Further Examples .......................................................................... 30

## 2.3 Advanced Set Theory     31

    Tuples .............................................................................................. 31
    Cartesian Products ........................................................................ 33
    Set Comprehension by Form ........................................................ 35
    Constructing Sets of Tuples .......................................................... 36
    Powersets ....................................................................................... 37

## 2.4 Venn Diagrams     38

|     | 2.5 The Model-Oriented Approach | 39 |
|---|---|---|
|     | 2.6 Summary | 40 |
|     | Key Topics in Set Theory | 41 |

# 3 LOGIC — 43

|     | 3.1 Predicate Logic | 44 |
|---|---|---|
|     | Other Predicates | 44 |
|     | Constraints | 45 |
|     | 3.2 Operations on Predicates | 46 |
|     | Equivalence | 46 |
|     | Negation | 47 |
|     | Disjunction | 48 |
|     | Truth Tables | 49 |
|     | Conjunction | 51 |
|     | Implication | 51 |
|     | Precedence Rules for Logical Expressions | 53 |
|     | 3.3 Quantification | 55 |
|     | Existential Quantification | 56 |
|     | Unique Quantification | 58 |
|     | Universal Quantification | 59 |
|     | 3.4 Summary | 62 |

# 4 BUILDING Z SPECIFICATIONS — 65

|     | 4.1 Objects in Z | 66 |
|---|---|---|
|     | Declarations | 67 |

## 4.2 Z Schemas 68
State Space .................................................................................. 72
Initial State ................................................................................. 75
Schema Reference......................................................................... 76
Changes of State .......................................................................... 78
Input and Output ........................................................................ 79
Δ and Ξ Schemas ........................................................................ 82
Combining Schemas..................................................................... 84
Scope ........................................................................................... 89
Axiomatic Descriptions................................................................ 90
Basic Types ................................................................................. 93

## 4.3 Specification Layout Conventions 94

## 4.4 Learning by Example 95
A Simple Computerised Library System...................................... 95

## 4.5 Summary 103

# 5 RELATIONS 105
Relational Operators ................................................................. 109

## 5.1 Fundamental Concepts 110
Domain ...................................................................................... 110
Range ......................................................................................... 111
Identity....................................................................................... 112
Relational Inversion .................................................................. 112
Relational Image ....................................................................... 113
Classification of Relations ........................................................ 115

## 5.2 Advanced Operations on Relations — 116
- Relational Composition .................................................. 116
- Iteration ............................................................................ 118
- Closures ........................................................................... 119

## 5.3 Manipulating Relations — 121
- Domain Restriction ........................................................ 121
- Range Restriction ........................................................... 122
- Relational Subtraction ................................................... 124

## 5.5 Summary — 125

# 6 FUNCTIONS — 127

## 6.1 Fundamental Concepts — 129
- Function Application ..................................................... 129
- Partial Functions ............................................................ 130
- Total Functions ............................................................... 132

## 6.2 Operations on Functions — 132
- Function Overriding ....................................................... 132
- Relational Operations .................................................... 134

## 6.3 Special Types of Function — 136
- Injections ......................................................................... 136
- Surjections ...................................................................... 137
- Bijections ......................................................................... 137

## 6.4 Summary — 138

# 7 SEQUENCES 141

## 7.1 Fundamental Concepts 142
Sequence Types .................................................................. 142
Indexing ............................................................................. 144
Length ................................................................................ 145

## 7.2 Operations on Sequences 146
Concatenation .................................................................... 146
Decomposition ................................................................... 147
Reversal .............................................................................. 149
Filtering .............................................................................. 150

## 7.3 Advanced Operations on Sequences 151
Distributed Concatenation ................................................ 151
Disjointness ....................................................................... 152
Partitions ........................................................................... 153

## 7.4 Summary 154

# 8 BAGS 155
Bag Types ........................................................................... 157

## 8.1 Fundamental Concepts 157
Bag Membership ................................................................ 157
Counting ............................................................................ 157
Creating Bags from Sequences ......................................... 158

## 8.2 Operations on Bags 159
Bag Union .......................................................................... 159
Bag Summary .................................................................... 160

| | | |
|---|---|---|
| 8.3 | The Z Data Type Pyramid | **160** |
| | Summary of Data Type Characteristics | 161 |

# 9 ADVANCED Z     163

| | | |
|---|---|---|
| 9.1 | Advanced Uses of Schemas | **164** |
| | Bindings | 164 |
| | Schema Types | 165 |
| | Selection | 167 |
| 9.2 | Miscellaneous Z Notation | **169** |
| | Generics | 169 |
| | Lambda Expressions | 172 |
| | More Schema Expressions | 174 |
| | Z Notation Not Covered | 174 |
| 9.3 | An Introduction to Proofs in Z | **174** |
| | Z : A Formal Notation | 175 |
| | Deduction and Proof in Z | 175 |

# APPENDIX A : CASE STUDY     179

# APPENDIX B : ANSWERS TO EXERCISES     195

# APPENDIX C : GLOSSARY     199

# INDEX     205

# Preface

Few doubt that formal methods will have a greater role to play in the next decade of Software Engineering. Governments are beginning to insist that software be formally specified; large firms with corporate clout are investing in formal know-how. In such a climate it is surprising that so few understand or even know of formal methodologies.

This book is an introduction to a formal language which was developed at the *Programming Research Group* at *Oxford University*. The language, called "Z" (and pronounced "zed"!), has been growing in popularity in both industry and academia. It is now the time to take a look at Z in a way that will help enlighten both these groups. This book endeavours to do just that.

The constructs of the Z Notation encourage specifications which are clear, well-presented, and intuitive. These are three very important properties which I believe no other formal methodology shares to such a great extent. Their importance lies in the doors they open for allowing more people to reap the benefits of formality in software development. It shouldn't be forgotten that many people have a fear of mathematics. The maths in Z is simple enough for a lot of these people to read; and while the more advanced aspects of proof are admittedly beyond their reach, this needn't be a problem since they can be insulated from such giberypokery.

Thus, this book serves as an easy-to-read, rather *informal* introduction to the fundamentals of Z. Software engineering practitioners will find this bias of particular importance. However, the book would also serve well as a text for a 1st or 2nd year undergraduate course in Software Engineering, Computer Science, or Applied Mathematics. With the growing importance being put on software engineering nowadays, the study of formal methods should be on every Computing syllabus.

## THE AIMS OF THIS BOOK

This book assumes no prior knowledge of any subject it covers, although a previous exposure to computing systems and advanced mathematics would of course be an advantage. However, some may flick through this book, see all the mathematical notation, and become a little apprehensive. DON'T PANIC! If an effort is made the notation can soon be mastered. It is a bit like learning to drive; some of the road signs and road markings of *The Highway Code* are rather strange at first, but one soon gains confidence with them.

- The book starts from *very first principles* in set theory and logic.
- There is a *step-by-step approach* to learning which gradually builds up knowledge, relating it to previous topics wherever possible.
- All the topics introduced are directly related to Z.
- The book covers that part of Z which is necessary for someone to be able to *read* Z specifications and understand the system described within them. Beyond this level, the book does not cover in much detail the additional material for someone wishing to, say, write Z specifications and *prove* them correct.

Other books teach set theory, relations, functions, logic etc. However, *An Introduction to Z* differs in that it provides a very gentle introduction to the topics. At all times, the relevance of the topics covered is clearly explained. Whenever possible, a direct use for the maths in a Z Specification will be given—illustrations will be given to show how each piece of notation can be used to form models of real-world systems.

# ACKNOWLEDGEMENTS

It was while I was an undergraduate at *University College London* that I first came across Z during a Software Engineering lecture. At the time, my peers remarked that Z was much more attractive than the other formal specification languages covered in the course, but that there was a lack of appropriate literature on Z. I told my tutor, Professor John Campbell, about their frustration. He came up with the simple answer: to write a book about Z with the requirement that it would be easily assimilated. This book is my response to his suggestion.

The source of this book was my third year Computer Science project. I am extremely grateful to my project supervisor, Dr. Maria Fox. She was very enthusiastic about the project, and spent a lot of her time reading and correcting the project manuscript. I sincerely thank her. I also thank my great friend Joanna Shallcross who gave me the all-important support while I was writing the book for the project, and who made the effort to meticulously study some of the chapters for errors in the re-written book later on.

Jim Woodcock sent me some interesting literature on Z, and pointed me in the direction of Dr. Mole of South Bank Polytechnic who provided me with some useful symbol fonts used in the book. I thank them both.

Finally, I would like to thank my family for their patience, with a special mention to my Grandmother who provided me with the Apple Macintosh this book was written with.

*Michael Imperato BSc (London)*

# CHAPTER ONE

# Introduction

In a world growing ever dependent on computers, there is a tremendous need for high quality software, and plenty of it. The fear that such a demand would far outstrip supply is what was often referred to as the *Software Crisis*. Computers are used in many places; from the mainframes of large governments, to the microchips in household toasters. These computers need software which can be produced using the minimum of resources, and which work as required. In some cases, it can be a matter of life or death; in areas such as defence, nuclear power, air-traffic control, and space exploration, we certainly cannot afford a malfunctioning computer.

Yet computer software development, as it is often practiced in industry today, is rather more of an *art* than a *science*. This was especially true in the early years of computing. As a creative process, it can of course lead to good computer systems, however not-so-good systems are also possible and as such the whole exercise can potentially be an expensive waste of time and money. Rife are the lurid horror stories spread around the industry of projects being millions of pounds over budget and many times over schedule.

A further complication (a consequence of the creative element of the process) is that there isn't even a universally accepted definition of what constitutes a 'good' computer system. Ostensibly, a good system should please the *users*. Considering the users' needs in software projects is a growing area of research, and includes the subject of *Human-Computer Interaction*[1]. An in-depth look at these albeit important matters is beyond the scope of this book. However, we will cover several aspects of user satisfaction later in this chapter when discussing the possible benefits of reading this book.

## Enter "Software Engineering"

*Software Engineering* is a discipline which evolved from the need to understand the process of developing automated information systems. It is an attempt to apply a scientific approach to software development in the hope of reducing the total effort involved, and increasing the likelihood of good software solutions produced in a more controlled manner. In short, software engineering aims to tackle the software crisis.

There are many approaches to software development which have surfaced over the years in software engineering. They all try to encourage a more structured approach with a systematic development strategy, each with a certain degree of success. However, none can be universally applied to all development projects—the project manager must select the methodology suited to the particular problem at hand.

Most methods produce satisfactory results when applied correctly to an appropriate problem. Many firms have their own unique refinements to existing methods. However, with so many around there is very little standardisation. Standards bring a degree of uniformity which is generally desirable. An established approach to specifying systems in a clear, unambiguous way would be a good thing.

---

[1] Human-Computer Interaction (HCI) is the study of human factors in computer systems. Many design strategies have emerged in the framework of HCI which aim to build computer systems around the users by considering their needs, rather than vice-versa. An interesting book on this subject is *Designing the User Interface* by B. Schneiderman (Addison-Wesley, 1987).

# The Communication Problem

Suppose a computer firm was contracted to produce a system for a client company. Very often the final computer system delivered is not exactly what the client had in mind. Systems are usually a compromise between what the client wanted and what the developer thought the client wanted. Indeed, it is not that rare for systems to be completely unusable as far as the client is concerned, in which case either a totally new system is required or the client goes back to the old way of doing things!

The main problem seems to stem from a lack of effective communication between the people involved. The more successful project normally arises from clear, precise communication throughout—where the client and developer both understand each others requirements, and the programmers know precisely what they must build. There has been much research within software engineering into making the successful project more commonplace.

When we *specify* a system, we are stating only *what* the system should do. This is a useful idea because it means we can concentrate on the functionality of the system without having to worry about the details of *how* we must design and code it. From a system specification document, designers and programmers can develop the system using any method they wish and are free to chose any computer configuration they wish. Of course, the idea is to produce an end system which does precisely what it was specified to do.

Imagine writing the specification of an information system using plain English prose. It is often extremely easy for the reader to misinterpret what is written simply because English is full of ambiguity, imprecisions, and innuendo. Yet, the system specification is a very important document. It can serve as a contract between the client and developer, and is often "signed-off" once accepted by both parties. A communication problem here means that everyone can have different views on what they think has to be done, a fact which sometimes only comes to light when the product is shown to the client or does not work quite as expected. These sorts of mistakes can be very costly to rectify.

Fortunately, the influence of software engineering means more and more specifications nowadays take into account the requirements for clarity. They use diagrammatic conventions or more structured forms of language. However, it is very hard to know that the client really understands them 100%, and even harder to be sure that the documents themselves are consistent and correct.

## 1.1 FORMAL SPECIFICATION LANGUAGES

Software engineering often uses theory from the sciences; *Computer Science* is one of the more obvious sources. A subject which has recently begun to emerge from the corridors of the universities and gain recognition in the world of software engineering is that of *formal specifications*. It often comes under the banner of *"formal methods"* since it makes use of the formality of Mathematics.

### What are Formal Specifications?

Mathematicians have their own language. They use various symbols to represent precise mathematical concepts. Thus, a string of such symbols has a definite meaning. It is a concise, unambiguous statement which often needs many lines of carefully selected English to be expressed with the same degree of exactness.

Formal specifications are system specification documents which use mathematical notation to precisely specify the attributes of an information system in an attempt to get rid of ambiguity. The documents are said to be written in a *formal specification language*. The mathematics in the language takes control of the sloppy English. A further very important advantage stems from the fact that it is relatively easy to rigorously *prove* properties of formally specified systems because of their mathematical foundation. We can use logical inference to prove that a system is consistent and will work as specified[2]. This is an impossible task when tried with computer programs written in languages whose semantics are not known.

---

[2] One area of research attempts to take a formal specification and add detail to it in a controlled way, in order to eventually end up with something that can directly be translated into a computer program. This is known as a *formal transformation*. This approach has lead to significant

Whilst some of the formal specification languages in existence can admittedly appear rather weird and not at first intuitive, others are really not so hard to understand. This book provides an informal introduction to one of the more approachable languages, called Z. Someone with a logical mind and the will to learn should have no real difficulties in understanding it.

## The Need for Formal Methods in Software Engineering

Formal methods rely on mathematics to provide a rigorous framework upon which we can model information systems. The mathematics allows us to take full control of the system model:

- it often provides an important insight into the problems we wish to model,
- we can get a better understanding of the underlying concepts of the system,
- it becomes possible to easily identify inconsistencies,
- the possibility for ambiguity is greatly reduced,
- it provides an apparatus for abstraction to help us cope with large computer systems,
- the level of control can allow us to make better project estimates of resources, and
- testers have access to a rich source for validation and verification.

The important point is that we can gain better control of what is a complex process. *Formal specifications* encourage a precise, unambiguous model of an information system, which has very important consequences:

■ An unambiguous model will be interpreted by different people in the same way. This means that we have a more reliable vehicle for communicating the specifications of a proposed computer system. The client and software developer both enter the design phase in the knowledge that the specifications accurately reflect what the final

---

reductions in development costs in some experiments carried out. The reason for this is that if we start from a correct specification and carry out correctness-preserving transformations, then the resulting program should also be correct.

system should do. The communication problem can be greatly reduced.

■ Members of the design, implementation, and testing teams will have an accurate *single reference point* to turn to, and one which is unambiguous. This reduces the chances of inconsistency in large projects (where it is common for problems when interfacing the separately produced components of the system). It also provides a useful reference for system maintenance.

It is possible to mathematically prove that a specification is consistent[3], and that a specification is complete i.e. it covers all possible eventualities. These are two important features—they can save a lot of time and money. Mistakes in a specification tend to be the costliest to rectify in the testing and maintenance stage.

## The Problems With Formal Specifications

Industry has not thrown away all other non-formal methods and flocked to take up formality. There are problems with formal specifications which have caused their take-up to be slow:

■ There is a lack of knowledge of the benefits to be gained from formal specifications. Some people in industry have never heard of them, others simply disregard them as trivial or unworkable. This is partly because people are relatively happy with the methods they currently use—changing to formal specifications is a big step which some are just not prepared to take.

■ The amount of available documentation on any particular formal method tends to be thin on the ground. This slows down, and even discourages, the take-up of the methods as people feel there is a lack of support.

---

[3] Most formal specification languages, including Z, are based to a large degree on first-order predicate logic. This represents what is generally believed to be the most powerful *complete* system in Mathematics. Its *soundness* means that we can find inconsistencies in specifications i.e. if we know a statement P is true, then the opposite of P cannot also be true. It allows us to reason about our specifications with the confidence that the logical apparatus that we are using is sound and generally well-formed.

- An important point is that software developers will gain little from 'unilaterally' deciding to use formal specifications. It is essential for the customer also to be able to understand the specification documents themselves. If the customer cannot understand the language then they cannot sign the document. This would mean that an English version of the document would need to be prepared, which gets rid of one of the important benefits of formal specifications—the idea of an accurate *single* reference point. Also, the English version signed could be ambiguous, inconsistent, and incomplete and appear to say something different to the formal version. The communication problem again rears its ugly face.
- A fear of mathematics is prevalent outside the sciences (and even within them!). People are put-off by the symbol manipulation, and the seemingly complex presentation of the mathematics.
- The need for abstraction in formal specifications is something which is hard to teach and practice. It is sometimes hard to decide exactly when to abstract away detail, and in what mathematical way to form any abstraction needed.

These problems are serious and need to be tackled if the full benefits of formal specifications are to be realised. Among other things, this book hopes to show that the mathematics needn't be difficult to learn, and that the client could quite painlessly learn to read formal specifications.

## Approaches to Formal Specifications

There are two approaches to formal specification languages which fundamentally affect the way in which a system is specified.

A *model-oriented* approach to specification aims to construct an explicit abstract model of an information system in terms of well-understood mathematical entities such as *sets* whose semantics are formally defined.

Z has such an approach to system specification. Thus, Z is said to be a *model-based specification language*. It is believed that specifications in such a language are more intuitive to the non-scientist since one models real-world entities directly using relatively simple mathematical objects. This means that the reader can more easily form a model of the system in their mind.

An *algebraic* approach to specification involves creating objects which represent some real world entities, and modelling them in terms of the operations which can be performed on them. Thus, we describe an information system in terms of its desired properties. There is no explicit model structure created. Such a formal methodology closely mimics an *object-oriented* approach to designing systems (the advantages of which are well-known nowadays). It is even possible to *animate* some algebraic specifications. This can be of tremendous value as a means of prototyping systems in a mathematically sound way.

The problem with algebraic specifications is that it is hard for non-scientists to get to grips with them. This is because the specification documents they produce are not really intuitive. As there is no explicit model of the system it is left to the reader to build it up mentally, and the equational nature of the specifications makes this very hard for the non-scientist.

It is important that the formal specification language which becomes widely used is easy for non-scientists to read, and yet still has the power to be able to model large, complicated information systems.

## 1.2 INTRODUCING Z

Z is a formal specification language devised by Jean-Raymond Abrial and developed by the *Programming Research Group* at *Oxford University* in the early '80s. It is still the subject of much research at Oxford and other institutions, and has been the centre of much interest from the non-academic world. Companies such as IBM, *Inmos*[4], and *British Telecom*, as well as the *Ministry of Defence* have all used Z to some degree.

---

[4]  A famous example of *Inmos* using Z was when developing their transputer. Refer to *Transputer Instruction Set - A Compiler Writer's Guide*, Prentice-Hall, 1988.

# The Mathematical Foundations Of Z

*Set theory* provides the mathematical foundation for Z. The set theory utilised by Z is a small subset (no pun intended) of what is a large area of mathematics. The area of the subject used by Z is well understood by Mathematicians, and there is a high probability that this area is well-formed and mathematically sound (we cannot prove this for certain). Indeed, one of the advantages of Z is that its semantics have been formally defined[5].

In Z Specifications, the mathematical notation (called the *Z Notation*) is used hand-in-hand with a more informal English description of the system. Each new concept is introduced in English and then clearly presented in a 'chunk' of Z Notation which is kept as small as possible. This helps to make the new concepts more manageable for the reader—there are fewer new things to take in at any one time, which gives the reader greater chance to organise his/her mental model of the system.

To specify in Z, one must be able to abstract away from the details and think of the higher level views of the system. Not every aspect of a system can be specified. For example, it is not really possible to specify interactive elements of a program such as screen displays and real-time keyboard responses. However, this is not too much of a problem; the ability to specify the behaviour of the underlying information system means that the part which is most often misunderstood, and hence profits most from specification, can be modelled in Z.

# Why bother learning Z?

There are a growing number of proponents of formal specifications. They say that such specifications can reduce software development costs and facilitate software maintenance (the process of improving programs, adapting programs to new requirements, and correcting errors in programs). Yet, the vast majority of people in information technology (IT) have either never heard of formal methods, dismiss them as theoretical and not applicable to

---

[5] Spivey, *Understanding Z*, Cambridge University Press, 1988.

the real world, or feel that they are far too complicated and best left to the academics.

Z has been used to specify several *non-trivial* information systems. It is generally used more for safety-critical projects at present, but that will hopefully change as more and more people hear about Z.

Of course, Z is of interest in academia because of its mathematical foundations, and because of the promise of being able to bring a degree of rigour to a software engineering project. However, some doubt whether this can be achieved universally. They are probably right, but that does not detract from the very real benefits, discussed in the previous section, that could accrue from using Z in the real world.

One of the biggest advantages of Z as a language for specifying medium to large-sized systems is its in-built *schema calculus*. This provides a mechanism for easily decomposing specifications into smaller, more manageable units called *schemas*.

Schemas give Z specifications a modular property, something which has long been recognised as a powerful aid when dealing with complicated problem domains. It helps the analyst to more easily build a correct specification, and allows the reader to be gently introduced to a new specification by gradually unravelling the model. It basically reflects the human shortcoming of only being able to deal with a few new concepts at any one moment in time.

## The Disadvantages of Z

Z works fine when modelling many information systems, but there are some for which it is not practical. For example, it is very hard to specify *concurrent* systems with the state of the Z Notation at present. Similarly, it is difficult to find ways of specifying *real-time* systems since Z is devoid of the notion of time.

In addition, the Z Notation contains a graphical convention with many symbols which are non-standard on most computer systems. This hinders the development of Z documentation. Some people resort to pen and paper, others use their initiative to varying degrees of success.

With time these disadvantages will be overcome, but there is still some debate as to whether Z will be universally applicable. Z is certainly not a panacea for software engineering, and we must remember that it provides only certain safeguards, and errors can creep in at other stages of the development process.

### The Future Of Z

Z has steadily evolved over a decade, and there is nothing to suggest that it will not do so in the '90s. Many large and important firms are actively involved in developing the methodology along with the more academic backing of Oxford PRG, ESPRIT, and SERC. All this means that many new and interesting developments are going on[6].

The growing popularity of Z, and the tremendous support that it has, ensures that it will have a definite role to play in the future of formal specifications.

# 1.3 FURTHER READING

This book serves as an introductory text; it is an informal tutorial containing everything needed to be able to understand most Z Specifications. It covers only those aspects of notation and mathematics which are directly relevant to understanding Z to this level.

Z is still growing, and several aspects of the methodology are still being researched. Thus, details about proof and transforming specifications into correct programs are not covered since it is felt that they are not stable

---

[6]  For example, a strategy for *animating* Z specifications is being developed. The general idea is to systematically convert Z specifications into *Prolog* which is then optimised for efficiency, and executed. This is achieved by using a library of Prolog rules based on the formal semantics of the language constructs of Z, and converting the specifications step-by-step using these rules. Other research is looking at different ways of specifying concurrency in Z.

enough at present. Other books will doubtless appear which cover such advanced topics when they have been standardised.

Those readers who wish to find out more about Z, and perhaps seriously consider using Z in their own specifications, will find the following two books almost essential reading:

- *The Z Notation: A Reference Manual* by J. M. Spivey (Prentice-Hall, 1989). This book is the definitive guide to the Z Notation. It provides type rules, descriptions, and formal definitions for all the standard notation in the Z language. It is not especially easy to understand, and a knowledge of BNF grammars is needed. Despite this, the book remains highly recommended.

- *Specification Case Studies* edited by I. Hayes (Prentice-Hall, 1987). This book provides several examples of the use of Z Specifications in real-world situations. It makes very interesting reading, and represents an excellent vehicle for getting well-acquainted with the Z Notation.

There are many other books which cover the mathematics used in software engineering. Most of them include material similar to that in this book, some also explore other areas. The following books are examples of the better books available:

- *Introduction to Discrete Mathematics for Software Engineering* by T. Denvir (Macmillan, 1986). An excellent book which starts from a very basic level and explains much of the material covered in this book from a slightly different angle and without the Z bias.

- *Software Engineering Mathematics* by J. Woodcock & M. Loomes (Pitman, 1988). This book covers some of the important aspects of Z with an emphasis on mathematical proof. Some may find this a little hard going, but it is an excellent book if you wish to take Z further. It also discusses some other formal methods.

# CHAPTER TWO

# Set Theory

We start this book with a subject most readers will have probably come across in their school days. The material contained within this chapter is without doubt comparatively elementary in nature. However, set theory is of fundamental importance when specifying systems in Z; it provides the foundation and the building blocks from which a model of a computer system may be built. Because we are using sets as tools in a specific way the emphasis will be different to standard school set theory. For this reason it is advisable for even the more confident reader to study this chapter. It is also important to become familiar with the notation used in Z. We will start with an informal definition of a set.

*Definition*

A **set** is a collection of objects. These objects must be definite and distinct. This means that each object in a set is different to all the other objects in the set, and the objects make sense—they each represent some discernible entity. Thus, given a set and an object we can see unequivocally whether or not that object is in the set.

### Example 2.1

Consider a bookshelf. It can hold a collection of books, each one being an object. We can treat the bookshelf as a set of books. No two books on the bookshelf are the same—the books are distinct.

In Z we are mostly interested in collections of *related* objects. For example, a collection consisting of a cat, a lamp, and an arrow is a set but the objects are all rather unrelated! In modelling real systems we often come across collections of objects which naturally relate e.g. the set of all employees in a firm (all the objects are employee names), or the set of rooms available in a hotel (all the objects are room numbers).

## An Introduction to Objects

At this point we should note that the term 'object' can refer to almost any entity in the real world that we are trying to model. In the previous example, the objects we dealt with were books. The bookshelf itself is an entity and so it is also an object; in this case we introduce the possibility of having **composite** objects. Such objects are composed of other objects. The bookshelf (a set of books) is an object composed of book objects.

Chapter four will explain in more detail how we deal with objects in Z. For now we will say that some objects are 'fixed' and others are 'variable'. The following are two examples of such objects:

- ▪ `Mark` is a fixed object since we wish it to represent a single specified real-world entity—a person in this case.
- ▪ *reader_name* is a variable object since, although it is a single object in Z, it can be set to represent one of billions of people.

The value of the first object is fixed to `Mark`, while the value of the second object could depend on many possible factors, and may vary during the lifetime of the computer system. Throughout this book we will use variable objects (**variables**) in most cases since they represent *concepts* or ideas (e.g. the concept of a car in general as opposed to a particular example of a car), and as such are much more flexible than fixed objects.

## Notation

Mathematicians love notation—they use weird and wonderful symbols whenever they can. Of course, they do have a good reason for this. It is based on the fact that one symbol can precisely and concisely convey a potentially large amount of information. This is *abstraction*—a mechanism for dealing with complexity which we will be coming across throughout this book. Don't be put off by the squiggly lines and dots; they are just shorthand which can soon be mastered.

In this book many new pieces of notation and terminology will be introduced. The broad strategy will be to introduce the new concepts in the text and then to illustrate their general form after a black star. By 'general form', it is meant that arbitrary objects will be used (which represent any object); these objects will be letters of the alphabet. All Z objects introduced will be printed in a distinctive typeface, e.g.

> *bookshelf*, `Macintosh`, *book*, $x$, $A$

Note that fixed object names will be printed in a typewriter-style face. Important new terminology will be highlighted in bold print when first introduced.

## Set Display

Sets have their own particular notation. We can **display** sets by listing all the objects in the set. We separate each object using commas, and surround the list in braces (curly brackets).

✱     $\{a, b, c\}$
       ...represents the set containing the objects $a$, $b$, and $c$.

In the above general form of a set display we called the objects $a$, $b$, and $c$ to keep things simple, but they could have been called anything. We choose simple letters for convenience and economy. In practice, Z objects have more meaningful names.

## Example 2.2

We will display the set of all colours in the rainbow:

{orange,blue,red,yellow,violet,green,indigo}

We can give a name to any object in a Z Specification. This is very useful when naming sets because the name *represents* its contents. Instead of having to write down precisely what is in the set each time we wish to refer to it (there may be a huge number of objects in the set), we need only write a name and the reader will know immediately which collection of objects we mean. Such information compression is a form of abstraction.

The names given to objects, be they fixed or variable, are known as **identifiers**. Usually in mathematics set identifiers are single capital letters like $S$ and $T$. However, Z Specifications should have longer descriptive names assigned to their sets. In this way it is easier and quicker to identify (and remember!) what a set contains which leads to clearer specifications.

## Example 2.3

Let's define the set of colours in the rainbow and name this set:

*rainbow* = {orange,blue,red,yellow,violet,green,indigo}

Notice that there is the name of the set on the lefthand side (LHS), then an equals sign, and then the set display on the righthand side (RHS) which shows what the identifier *rainbow* refers to.

## Example 2.4

As a further example, let us define the set of ID numbers of the people allowed to pass through a security barrier:

*clearance* = {23312, 29884, 16778, 34127, 66931}

The naming convention used above is really like asserting an equivalence; the word *clearance* is equivalent to everything on the RHS, i.e. {23312, etc...}.

# 2.1 FUNDAMENTAL CONCEPTS IN SET THEORY

Now that we know what a set is, we can introduce some fundamental properties of sets and show the notation used in Z to represent these properties.

## Membership

Every object in a set is called a **member** or an **element** of that set. Given a set and an object we can say whether or not that object is a "member of" the set or, equivalently, whether or not it "belongs to" the set.

### Example 2.5

The ID number 23312 is a member of the set *clearance*. Similarly red is a member of the *rainbow* set, but brown is not a member of the set.

We shall now introduce our first strange-looking mathematical symbol (based on the Greek epsilon). It represents the concept of set membership:

*     $x \in A$
  ...means the object $x$ is a member of the set $A$.

*     $x \notin A$
  ...means the object $x$ is not a member of the set $A$.

The epsilon symbol is known as the membership **operator**. An operator is a mathematical symbol which represents the application of a mathematical operation to one or more objects (referred to as **operands**).

An operator which we are more familiar with is the arithmetic addition operator, the plus symbol "+", which is applied to numerical operands. The addition operator signals an addition operation. The result of such an operation is a number. Examine the following *arithmetic expression* built from variables:

$b + c$

i.e. add $b$ to $c$. Suppose the result is $a$. Clearly, $a$ could be one of an infinity of numbers. Now look at the following expression:

$b \in C$

i.e. $b$ is a member of $C$. Suppose the result is $P$, then $P$ could only take one of two possible values: *true* or *false*. It is true if $b$ is in set $C$. It is false if $b$ isn't in $C$. A Z expression whose value is taken from the set {*true, false*} is sometimes referred to as a **predicate**.

### Example 2.6

With the examples we have been using up to now, we can see that the following statements are true:

- green ∈ *rainbow*
- 39009 ∉ *clearance*

Note that in the above example the statements are true. Generally in Z we only want to state true facts (we can *assert* something to be true in a specification), but we must bear in mind that such statements can also be false. We will discuss this further in the next chapter which is about logic.

## Equality

We say that two sets are **equal** if, and only if, they contain precisely the same objects—no more and no less.

* $A = B$
  ...denotes the predicate which is true when set $A$ is equal to $B$.

* $A \neq B$
  ...denotes the predicate which is true when set $A$ is not equal to set $B$.

Notice that the equals sign here *appears* to have a different meaning to that used earlier on. Previously it was shown to be used to assign a name to an object. In equality we are using the equals sign as a predicate saying that the LHS and the RHS contain the same objects (which can be true or false). However, if we remember the convention that Z predicates can be asserted to be true, then giving a name to an object using the equality operator is like saying that the object on the LHS (a newly introduced identifier) contains the same information as those objects on the RHS—that they will be equal from now on.

**Example 2.7**

The following are true predicates:

- *clearance* = {23312, 29884, 16778, 34127, 66931}
- *rainbow* ≠ {red, amber, green}
- {1, 2, 3} ≠ {1, 2, 3, 4}
- {1, 2} ≠ {1, 3}
- {4, 1, 2} = {1, 2, 4}

The fifth example highlights an important property of sets. Sets are **unordered**. By this we mean that the order in which we write down the members of sets is unimportant. A set does not contain any ordering information. Thus a set is like a sack which contains several different objects; we are cannot possibly know the precise location of an object in the sack, we simply have the ability to know whether or not the particular object is in the sack.

## Cardinality

A final thing we can say about a set is the number of members it has, or equivalently, the number of objects it contains. For historical reasons this number is known as the **cardinality** of the set. In Z we denote the cardinality with the number symbol.

* #$A$
    ...denotes the number of elements in the set $A$.

### Example 2.8

The following are true predicates:

- $\#rainbow = 7$
- $\#\{4\} = 1$

A set having a cardinality of one is known as a **singleton set**. Note that the objects $x$ and $\{x\}$ are different. The former represents a simple object called $x$, whereas the latter represents a singleton set which happens to contain $x$.

*Discussion*

In Example 2.1 we were careful to state that no two books in the set were the same. When we model real-world data we may come across repetitions. For example, if we wanted to model a payroll system we could consider using a set to hold the names of all the employees. However, if two employees had the same name then there would be no way of differentiating between the two using the basic sets covered so far.

Thus, there is no such thing as a set $\{a, a, b, b, a, b\}$ in Z; the set is $\{a, b\}$, and has only two members. We can test to see if $a$ is a member of the set, but finding the number of occurrences (greater than one) of $a$ in the set is meaningless—by definition a set's members are all unique. There are many ways of dealing with repetitions in Z Specifications, and they will be discussed in later chapters. (We could still use a set in the payroll example by assigning a unique I.D. to each employee).

## The Empty Set

The **empty set** is a special example of a set. It is the set which has no members. So by definition, no object may ever belong to the empty set.

✱  $\{\,\}$ and $\varnothing$
   ...both symbols denote the empty set.

## Exercise 2.1

What is the cardinality of the empty set?

## Set Inclusion

The objects which a set contains can be sets themselves. This is because a set is just another object—a composite object. This is very useful because it allows us to build up hierarchical set structures, which in turn facilitates the construction of hierarchical models of forms that often crop up in the real world.

While a set may legally contain members which are sets, each set within the set constitutes a single member. For example, the set $A = \{3, 4, B\}$ has only three members—a fact which is not affected by the contents of the set $B$.

## Exercise 2.2

Suppose we have the sets

$A = \{2, 4, \{\}, 8\}$

$B = \{10, 20, A, 40\}$

Which of the following predicates are true?

a) $\#A = 4$
b) $\#B = 7$
c) $10 \notin B$
d) $\{4\} \in A$
e) $\{2, 4, \{\}, 8\} \in B$

When we define a set we must be careful about self-reference. This is when a set has a reference to itself within its own definition, and is not allowed.

## Example 2.9

The following is an illegal definition of a set because it attempts to define a set in terms of itself:

$$A = \{A, B, C\}$$

If we tried to write out the set in full there could never be enough paper to finish the set display since it is infinite in length:

$$A = \{A, B, C\} = \{\{A, B, C\}, B, C\} = \{\{\{A, B, C\}, B, C\}, B, C\} = \text{etc...}$$

## Subsets

Just as we can test whether a certain object is a member of a set we can also test whether a certain set of objects is contained within another set, i.e. whether it is a **subset** or not. A set $A$ is said to be a subset of another set $B$ if all the members of $A$ are also members of $B$.

*     $A \subseteq B$
    ...denotes the predicate which is true when set $B$ is a subset of set $A$.

By definition, if $A = B$ then $A \subseteq B$.

*     $A \not\subseteq B$
    ...denotes the predicate which is true when set $A$ is not a subset of the set $B$.

Of course, if $A \subseteq B$ and $B \subseteq A$ then $A = B$.

The empty set is assumed to be a subset of all possible sets (including itself).

The above notation is the most commonly used when using sets. However, there are some occasions when we wish to emphasise that one set $A$ is a **proper subset** of another set $B$, i.e. all the members of $A$ are members of $B$ but $B$ has at least one other member which is not in $A$.

*     $A \subset B$
    ...denotes the predicate which is true when set $A$ is a proper subset of set $B$, and hence where $A \neq B$.

∗   $A \not\subset B$
   ...denotes the predicate which is true when set $A$ is not a proper subset of set $B$.

### Exercise 2.3

Decide which of the following predicates are true:

a)  $\{a, b\} \subseteq \{c, a, d, f, b\}$
b)  $\{x\} \subseteq \{x\}$
c)  $\{\text{red}, \text{mauve}, \text{blue}\} \subseteq \textit{rainbow}$
d)  $\{\{\}\} \subseteq \{\{\{\}\}\}$

## Finite Sets

A **finite** set $F$ is one which contains a *bounded* number of members, i.e. one for which we can say $\#F = n$ where $n$ is a positive number. In Z we will primarily be concerned with such finite sets since real-world systems are finite; that is, there is generally an upper bound on the number of objects we wish to manipulate. For example, a government database which holds information on the entire population may be huge, but it is certainly not infinite.

## Special Sets

There are some special sets used in Z Specifications which are very important and are frequently utilised:

■   The set of **Integers** is the set of all whole numbers, i.e. it is the set

   $\{..., -2, -1, 0, 1, 2, 3, ...\}$.

   Note the use of the ellipsis to indicate that the numbers continue on to infinity in the positive and negative directions. So, for example, 5,778 is a member of the set of integers, as is –67,234,567,120.

■   The set of **Natural Numbers** is the set of all nonnegative integers.

* **Z**
    ...denotes the set of all integers. Hence, **Z** = {..., -2, -1, 0, 1, 2, 3, ...}.

* **N**
    ...denotes the set of all natural numbers. Hence, **N** = {0, 1, 2, 3,...}.

Clearly **N** ⊆ **Z**. We tend to use the natural numbers more than any other set of numbers in Z Specifications. Both **Z** and **N** are examples of **infinite** sets.

Quite often it is convenient to work with a **number range**, i.e. a subset of **Z** or **N** which contains all the numbers in a given range. For this there is a special piece of Z Notation:

* $x..y$
    ...denotes the set of integers between $x$ and $y$ inclusively.

By definition, if $x$ is greater than $y$, then the expression $x..y$ is equal to $\emptyset$.

### Example 2.10

The following are examples of the number range notation:

- $1..6 = \{1, 2, 3, 4, 5, 6\}$
- $-4..2 = \{-4, -3, -2, -1, 0, 1, 2\}$
- $7..7 = \{7\}$
- $7..6 = \emptyset$

## Set Comprehension

The method we have been using for defining and displaying sets up until now is fine for those containing just a few objects, but when the list is large it becomes impractical. For this reason we will introduce a notation to define a set using a method other than explicitly writing out all of its members.

When we describe a large set in English we often say it contains all those objects with a certain property, i.e. we can tell if an object is a member of a set simply by checking that the object has the required property. For example, we may say that a set contains all those employees who have had at least ten years experience in some field (this is the property we wish to assert). If an employee has had less than ten years in that field then he or she is not in the set.

Some examples of predicates which we have met so far are $x \in A$, $A \subseteq B$, and $A \neq B$. All three of these can either be true or false depending on the contents of the sets $A$ and $B$. We will use predicates in an implicit form of set construction known as **set comprehension**.

*     $\{x : T \mid P\}$
     ...denotes the set of all $x$ of type $T$ such that $P$ is true.

Forget for a moment that the above notation mentions the word 'type'. What it does is implicitly describe the set of all objects $x$ for which the predicate $P$ is true. Take any object and call it $x$. If $P$ is true given this $x$ then $x$ is a member of the set, otherwise $x$ is not a member of the set. For example, $P$ could be the predicate "$x$ is greater than 5" in which case *if* the $x$ we supply *is* greater than 5, then that $x$ is a member of the set. "6" could be a member of the set, while "4" couldn't.

We have glossed over an important point. In the above example we are assuming that $x$ is a number so that "$x$ is greater than 5" makes sense. However, say $x$ was the colour purple; "`purple > 5`" doesn't really make much sense at all! This leads us to the notion of a **type**. A type is a set. Every object in Z is associated with a type. If two objects are of the same type, then they are both drawn from (are members of) the same set. For example, the set of natural numbers **N** can be thought of as a type, and when we introduce an object $x$ which is a natural number we know $x \in$ **N**, and say $x$ is of type **N**.

It is important to note that all the members of a set in the Z Notation must be of the same type. Set comprehension ensures this by specifying the type after a colon following the identifier. However, when we construct sets using the simple explicit set display used earlier, we must ensure that all members are of the same type ourselves.

## Example 2.11

Let's say that we have a set called *colour* which contains the names of all the colours discernible to the naked eye. We will construct *rainbow* in a different way:

$rainbow = \{x : colour \mid x \text{ is a colour in the rainbow}\}$

Here *rainbow* is described as a set which contains colours, and these colours are colours of the rainbow.

Clearly, the RHS of the above assignment is exactly equivalent to the RHS of the assignment in Example 2.3.

## Example 2.12

Consider the set of all even natural numbers less than 100. We can define the set as

$even = \{x : \mathbb{N} \mid x \text{ is less than 100 and } (x \bmod 2 = 0)\}$.

Here the 'evenness' of the number is asserted by taking the remainder when dividing the number by two (i.e. $x$ modulo 2); the number can only be a member of the set if the remainder after division is zero *and* the number is less than 100.

Note that the words "is less than" and "and" are not part of the Z Notation; they were used here for clarity. The next chapter will introduce the correct mathematical notation for these concepts.

An important advantage of set comprehension is that it does not clutter the document and our minds with lists of objects. Instead, it gives us rules and concepts so that we can instantly visualise the contents of the set without having to scan a list and extract meaning ourselves. If we wish to know that a certain object is in the set then we simply check if the object is of the correct type and that it satisfies the predicate (i.e. we check if the predicate is true when we consider the object). This is yet another example of abstraction in Z.

# Examples

We will now consider some more examples of using sets, but this time with more emphasis on modelling real-world objects in a manner which is common to Z Specifications.

## Example 2.13   SHOPPING CATALOGUE

A database, in its simplest form, holds various groups of related items of data and thus can often be modelled with sets. For example, we could consider a shopping catalogue. It contains many items each with their own unique catalogue number.

Using set theory to help us we could model an enquiry system at a distribution depot which can say whether or not an item is in stock, and whether or not the item is on order. To do this we could have two sets; *instock* which contains the catalogue number of all items currently in stock, and *onorder* which contains the catalogue number of all items currently on order. All catalogue numbers are of type $CATNUM$, i.e. they are taken from the set $CATNUM$ which contains all possible legal catalogue numbers. An enquiry is simply a matter of checking whether the item in question is a member of *instock* and, if not, whether it is a member of *onorder*.

## Example 2.14   TIMETABLING

A University department offers its students a large list of possible courses. The student must pick four of these. However, due to the limited number of staff and lecture theatres, some course combinations will clash. We could model the group of all possible legal course combinations as a set which contains many sets of cardinality four (sets which contain four members).

For example, if each course has a unique code, of type $COURSECODE$, then the following is a possibility for the first few members of the set of all legal combinations:

$legalcombin = \{\{A3, B2, B6, D6\}, \{A2, A3, B1, C6\}, \{A1, A2, B2, D5\}, \text{etc...}\}$

From this set it is a simple matter to check whether a particular combination of four courses is supportable—just check for membership.

## 2.2 SET MANIPULATION

So far we have only discussed the static elements of set theory. We can define a set, decide whether or not an object is in a set, and count the number of members in a set, but as yet we have no operations which can be applied to sets to give new sets changed in some way e.g. with a member added or taken away. We will now discuss a more dynamic approach by introducing set operators which take existing sets as operands and produce modified sets as results.

### Set Union

Given two sets $A$ and $B$ of type $T$, the **union** of $A$ and $B$ is defined as the set which contains all those objects which are members of $A$ and/or are members of $B$.

✱  $A \cup B$
   ...denotes the set which is the union of $A$ and $B$.

Using set comprehension we can define $A \cup B$ more formally as

$\{x : T \mid x \in A \text{ and/or } x \in B\}$.

Note that the resulting set has a type identical to the type of the operands. This is not unreasonable since it would be meaningless if we added, say, two sets of employee names and got a set of colours of the rainbow! Similarly, $A$ and $B$ must be of the same type otherwise their union will not be a legal Z set (which can be of only one type).

Set union can be used on many sets at once, not just two.

### Example 2.15

If we had four sets containing the names of employees in the four different departments of a firm, call them $A$, $B$, $C$, and $D$, then the union of these sets would be the set containing the names of all the employees in that firm:

$employees = A \cup B \cup C \cup D$

Set union is thus a useful way of joining sets together.

## Set Intersection

Given two sets $A$ and $B$ of type $T$, the **intersection** of $A$ and $B$ is defined as the set which contains all those objects which are members of $A$ and are also members of $B$.

* $A \cap B$
  ...denotes the set which is the intersection of $A$ and $B$.

By definition,

$A \cap B = \{x : T \mid x \in A \text{ and } x \in B\}.$

An important observation that can be made is that $A \cap B = B \cap A$. Similarly, $A \cup B = B \cup A$.

Set intersection, like set union, can be used on many sets at once, not just two. Thus it is a useful way of finding the members that sets have in common.

## Set Difference

Given a set $A$ and another set $B$ of type $T$, we define $A \setminus B$ to be the set whose objects are members of $A$ and are *not* members of $B$.

* $A \setminus B$
  ...denotes the **set difference** of $A$ with respect to $B$.

By definition,

$A \setminus B = \{x : T \mid x \in A \text{ and } x \notin B\}$.

In some notations a minus sign is used instead of a backslash, i.e. $A - B$ is equivalent to $A \setminus B$.

It is important to note that $A \setminus B$ is not the same as to $B \setminus A$, and the resulting sets are very likely to be different. This is analogous to the arithmetic subtraction operation where $x - y$ is generally different to $y - x$. For example, $6 - 2 \neq 2 - 6$.

*Discussion*

The above three operators $\cup, \cap$, and $\setminus$ are only defined if their operands are sets of the same type. For example, if we have a set $A$ from which we wish to remove the object $y$ which is a member of $A$, then $A \setminus y$ is not defined. In this case we create a temporary singleton set $\{y\}$ and use $A \setminus \{y\}$ to achieve the desired result.

## Further Examples

With the set operations $\cup, \cap$, and $\setminus$ we have the basic tools for manipulating sets which can be used to mimic operations carried out in the systems we are trying to model.

### Example 2.16    SHOPPING CATALOGUE continued...

If we consider the shopping catalogue example in the previous section we can use the set operations to model the following operations:

- *OutOfStock* – an operation which deletes the code of an item which is no longer in stock, and
- *Arrived* – an operation which adds the code of an item, previously on order, which has arrived at the depot.

where each code is a member of *CATNUM*.

The first operation, *OutOfStock*, is simply a matter of removing a code from *instock*. If the code to be removed is $x$, then the updated version of *instock* will be equal to *instock* \ $\{x\}$.

*Arrived* involves adding a code to *instock* and removing the same code from *onorder*. If the code is $x$, then the new contents of *instock* and *onorder* will be *instock* $\cup \{x\}$ and *onorder* \ $\{x\}$ respectively.

---

The above example highlights an important concept in Z. The contents of *instock* before the *OutOfStock* operation are different from its contents after the operation. In Z we say that there has been a "change in state"—the value of an object (here the contents of the set) changes, and thus moves from being in one state to being in another. We will discuss the concept of state in more detail in Chapter Four.

For now we will introduce a convention in the Z Notation. The before and after states of an object can be distinguished by a **decoration** (which is a suffix added to the identifier). When an object name is decorated with a dash (´) it represents the final state of that object after an operation. Note that the type of an object cannot be changed by an operation. So, in Example 2.16 we could have written the following for the *OutOfStock* operation:

$instock´ = instock \setminus \{x\}$

which specifies that the operation results in the contents of *instock* changing to its previous contents minus the $x$ code.

## 2.3 ADVANCED SET THEORY

### Tuples

Sometimes we will want to deal with objects that are quite complex in nature. An object could have several properties, or be composed of other objects.

### Example 2.17  LIBRARY BOOK

If we had an object to model a library book then this object could have the following properties:

  i. The title of the book
  ii. The name of the author
  iii. The name of the publishers
  iv. The year of publication
  v. The ISBN of the book

We would like to have some way of being able to treat each book as a single object, and a way of extracting certain information about the books when required.

Representing each book as a set of the five properties is not what we want. For example,

    {The_Z_Notation,Spivey,Prentice-Hall,1989,0139-83768-X}

could represent a book. However, given such a set, how do we find out the year of publication for example? It is not possible using basic sets—we cannot ask for the fourth member because there is no such thing as a fourth member of a set; sets are unordered. The above example could easily have been displayed as

    {1989,Spivey,0139-83768-X,Prentice-Hall,The_Z_Notation}

Both sets are equal because they contain exactly the same members and are both of the same cardinality.

What we need is some method of creating objects which contain information that can easily be extracted.

*Definition*

A **tuple** is an ordered collection of objects. Each object may be of a different type. It is like an ordered set, but of possibly unrelated objects. A tuple which contains two elements is called an **ordered pair**. Tuples containing three elements are called 3-tuples, and, in general, tuples containing $n$ elements are called *n*-tuples. Elements of a tuple must be displayed in the correct order, separated by commas, and enclosed in brackets:

* (*a*, *b*)
  ...denotes an ordered pair with first element *a* and second element *b*.

* (*a*, *b*, *c*)
  ...denotes a 3-tuple.

## Cartesian Products

The **cartesian product** of a list of *n* sets is the set of all *n*-tuples formed from the sets in list order.

* *A* × *B*
  ...denotes the set of all possible ordered pairs (*a*, *b*) such that the first element is a member of *A* and the second element is a member of *B*.

* *A* × *B* × *C*
  ...denotes the set of all possible 3-tuples (*a*, *b*, *c*) such that the third element is a member of *C* etc...

Cartesian products go hand-in-hand with tuples. Both represent a very important method of creating complicated set structures and they are often used in Z Specifications. However, they are not very easy to grasp first time so there follows a few examples which will hopefully remove any doubts.

### Example 2.18    LIBRARY BOOK continued...

We are now in a position to model a set of objects which represent books and have the five properties expressed in Example 2.17.

We will create a set called *book* which contains all the possible books in the world. *book* will be a set of 5-tuples defined as

$$book = title \times author \times publisher \times year \times isbn$$

where *title* is the set of all possible book titles, *author* is the set of all possible book authors, *publisher* is the set of all possible book publishers, *year* is the set of all years, and *isbn* is the set of all possible legal ISBN codes.

We can see that any book in existence is a member of the set *book*. If we were given a book called $x$ (a 5-tuple) such that $x \in book$ then we could tell what year it was published by looking at it's fourth element.

The cartesian product described is infinite! However, in an abstract specification this is not a problem. If we later refined the specification we could limit the theoretical size of the set by restricting the component sets e.g. by saying that the title cannot exceed 50 characters, and that the year must be between 1600 and 2100 etc. Whatever we did, it must be remembered that the computer system need not store all the possible values *book* can take on, but just needs to be able to check that any value it does take on is legal.

## Example 2.19

Most computer displays are made up of thousands of tiny dots called "pixels". On a monochrome screen, the dots are either one colour or another e.g. black or white. The shapes we see on the screen are made up from these pixels.

Imagine we wanted to model a simple drawing program which would allow you to draw black lines on a white screen, and then re-position the lines later if you wish. We can specify a particular position on the screen by supplying the "coordinates" of the corresponding pixel. The coordinates are two values; the x-coordinate specifies the horizontal distance from the bottom left of the screen and the y-coordinate specifies the vertical distance from the bottom-left of the screen.

We can give the coordinates of any pixel on the screen as an ordered pair $(x, y)$. Assume the coordinates of the bottom-left of the screen are $(0, 0)$. A Line can be specified by providing two sets of coordinates representing the end-points between which the line may be drawn.

Our model could have a set called *lines* which is initially empty. For each line drawn, an associated set containing two ordered pairs will be in *lines*. When we wish to change the position of a line we simply identify the line in question and change the coordinates of the two ordered pairs appropriately. For example, to draw a new line from (45, 23) to (134, 99) we assert

$$lines' = lines \cup \{\{(45, 23), (134, 99)\}\}$$

## Set Comprehension by Form

There is another way of constructing sets implicitly. Instead of supplying the rules which define whether or not an object is a member of a set we can describe the general 'form' or 'shape' of the objects which are its members. Thus, to test whether an object is in a set one simply checks to see if the object is of the required form. This notation is known as **set comprehension by form**.

✳   $\{x : T \bullet k\}$
    ...denotes the set of all objects $x$ of type $T$ which can be written in form $k$.

### Example 2.20

The following are sets constructed using set comprehension by form:

- $\{x : \mathbb{N} \bullet x^2\}$ which is equivalent to $\{0, 1, 4, 9, 16, 25, 36, ...\}$

- $\{y : \mathbb{Z} \bullet 3y + 2\}$ which is equivalent to $\{..., -4, -1, 2, 5, 8, ...\}$

The general idea is that we apply the form to every member of the type to give the required set. In the first example this involved taking each member of $\mathbb{N}$, i.e. 0, then 1, then 2, etc... and calculating the square of each number. This builds a resulting set.

A final version of implicit set notation combines the first two methods we know to give the most general form of set comprehension.

✳   $\{x : T \mid P \bullet k\}$
    ...denotes the set of all objects $x$ of type $T$ which are of form $k$ when $P$ is true.

# Constructing Sets of Tuples

The notation we have covered so far for defining sets does not yet enable us to easily create sets of tuples. The mechanism for achieving this in Z is to declare the types of the objects in the tuple in the correct order, separated by semicolons, in the declaration part of the set comprehension.

* $\{x:A; y:B \mid P \bullet k\}$
  ...denotes the set of all ordered pairs $(x, y)$ of type $A \times B$ which are of form $k$ when $P$ is true.

Notice that here we have used a cartesian product to show the type of a set. It is clear that any possible ordered pair $(x, y)$ in the resulting set will be a member of the set $A \times B$.

## Example 2.21

Suppose we were modelling a simple accountancy program. A firm wants to be able to list all those clients who owe more than £500. The set *debtors* contains all those clients who owe the firm money, how much they owe, and a reference to the transaction in the Sales Ledger. It is a set of 3-tuples whose first element is taken from the set of all the firm's clients, called *client*; whose second element is the sum of money owed in pence; and whose third element is a cross-reference code which is a member of the set *ref*.

We can model those who owe more than £500 by constructing a set

$\{x:client; y:\mathbb{N}; z:ref \mid (x,y,z) \in debtors \text{ and } y \text{ is greater than } 50000\}$

where each entry is of type *client* $\times \mathbb{N} \times$ *ref*.

## Example 2.22

Suppose we had a set $T$ of ordered pairs $(x, y)$ where $x$ is of type $A$ and $y$ is of type $B$ and we wanted to take that set and create a new set containing exactly the same pairs but with the elements swapped, i.e. the member $(x, y)$ becomes $(y, x)$. We can achieve this with the following general form of set comprehension:

$$reverse\_of\_T = \{y:B; x:A \mid (x,y) \in T \bullet (y,x)\}$$

The members of *reverse_of_T* are of type $B \times A$, as opposed to $T$ which has members of type $A \times B$.

## Powersets

The **powerset** of a set $A$ is the set comprising all the subsets of $A$ including the empty set and $A$ itself.

* **P**$A$
    ...denotes the powerset of $A$.

### Example 2.23

The powerset of $\{1, 2, 3\}$ is denoted by **P**$\{1, 2, 3\}$ and is equal to the set

$$\{\{\}, \{1\}, \{2\}, \{3\}, \{1, 2\}, \{2, 3\}, \{1, 3\}, \{1, 2, 3\}\}.$$

Powersets are very useful when dealing with types in Z. Suppose that we have a set $A$ whose members are drawn from the set $B$, i.e. each member of $A$ is of type $B$. Every object in Z has a type, so the set $A$ must have a type and hence $A$ must be drawn from another set. This set is in fact **P**$B$, and so we say the type of set $A$ is **P**$B$. This may be hard to grasp at first; think that at any one moment in time $A$ contains one or more objects of taken from the set $B$, i.e. it is a subset of $B$. Now **P**$B$ is the set of all subsets of $B$. $A$ must be one of these subsets, and so $A \in$ **P**$B$ is always true.

### Example 2.24

Let $B$ = {Paul,Shams,Charlotte}. Then

**P**$B$ = {{}, {Paul}, {Shams}, {Charlotte}, {Paul,Shams},
    {Shams,Charlotte}, {Paul,Charlotte},
    {Paul,Shams,Charlotte}}.

Now, let $A = \{x : B \mid x \text{ is male}\}$. So, $A = \{\text{Paul, Shams}\}$, which is a member of $\mathbb{P}B$. Since $\mathbb{P}B$ contains all the possible sets which are constructed by taking members from $B$ it is plain that $A$ (which by definition will be a subset of $B$) will always be a member of $\mathbb{P}B$, hence it has type $\mathbb{P}B$.

### Example 2.25   TIMETABLING continued...

In the University timetable model of Example 2.14 we specified the set *legalcombin* which contained all legal course code combinations. Each course code was of type *COURSECODE*. *legalcombin* was a set containing sets of four course codes, and so was of type $\mathbb{P}(\mathbb{P}COURSECODE)$.

## 2.4 VENN DIAGRAMS

It is sometimes clearer to represent sets diagrammatically. We can use **Venn diagrams** (named after the Mathematician John Venn) to picture sets. Their use is not required in the Z Notation where specifications are written using the symbolic notation we have described up until now.

Venn diagrams represent sets as ovals. All points within the oval are members of the set, all points outside the oval are not members of the set. Figure 2.1 shows the sets $A = \{a, b\}$ and $B = \{b, c\}$ represented as two ovals containing labelled points. Note that $b$ is a member of both sets, and $d$ is a member of neither.

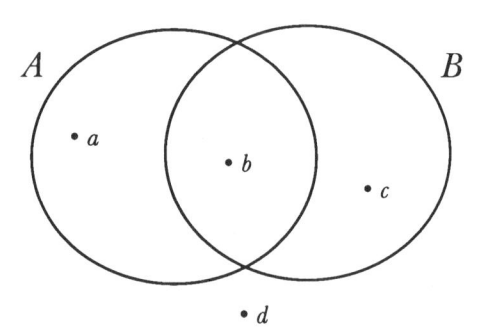

**Figure 2.1**

We can see from the diagram that $A \cup B$ is all the points within the two ovals, i.e. $\{a, b, c\}$, and $A \cap B$ is all the points within the overlap of the two ovals, i.e. $\{b\}$. Figure 2.2 shows $A \cup B$, $A \cap B$ and $A \setminus B$, for any two sets $A$ and $B$. Of course, not all pairs of sets have members in common, and so not all pairs of ovals will overlap in Venn diagrams.

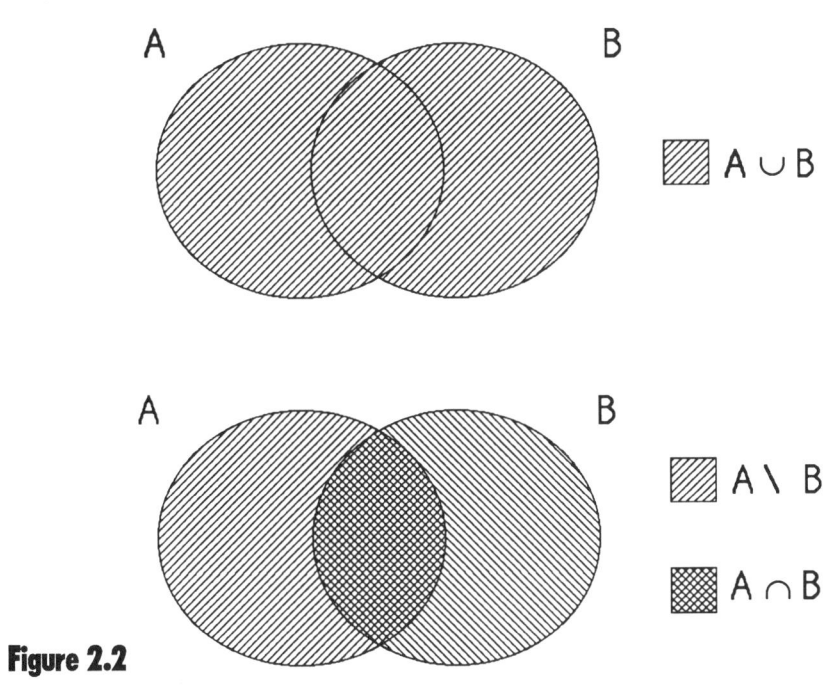

**Figure 2.2**

## 2.5 THE MODEL-ORIENTED APPROACH

It has been stated that Z uses set theory, but in what way does it use such a relatively simple calculus to specify elaborate real-world information systems? The answer is that a Z specification builds an explicit model of a system using set theory to develop an abstract mathematical model of the possible *states* the system can be in, along with mathematical descriptions (or specifications) of the operations which can be performed to transform one state of the system to another. This is called a *model-oriented approach* to formal system specification.

Viewing an explicit model of a system, built from the rich set of mathematical data types provided in the Z notation, is something which many people are capable of doing. The hardest part is probably getting to know the data types in the first place, and it has been argued that this should not be too much of a problem.

The advantage is that the reader of a specification is given all the information needed to be able to easily build a correct mental image of the system, in whole or in part. This mental image can then be enriched by the operation specifications which can be readily related to the mental model. The concept of state, and that a system can be viewed as a set of states, is appealing since it represents a way of somehow bringing order to what can be thought of as a intricate system.

# 2.6 SUMMARY

In this chapter we introduced the concept of a set as a collection of objects. From a given set, we can only find out if an object is a member of it—we cannot find out anything else. There is no concept of order in a set, and no concept of repetition—it doesn't make sense to say that a set contains $n$ occurrences of the same member, or that member $x$ comes before member $y$ etc. Cartesian products provide a way of introducing ordered objects.

We covered different mechanisms for describing sets. All the information about a set is contained within curly braces. Explicit notation shows all the members of a set separated by commas, whereas set comprehension defines the 'rules' which decide whether an object is a member of the set and provides a more implicit notation useful for abstraction.

The diagram overleaf provides an overview of the key topics covered in this chapter.

The Z Notation is based around set theory. If you understand sets and are completely at ease with using them, then the other aspects of Z should be much easier to grasp.

# Key Topics in Set Theory

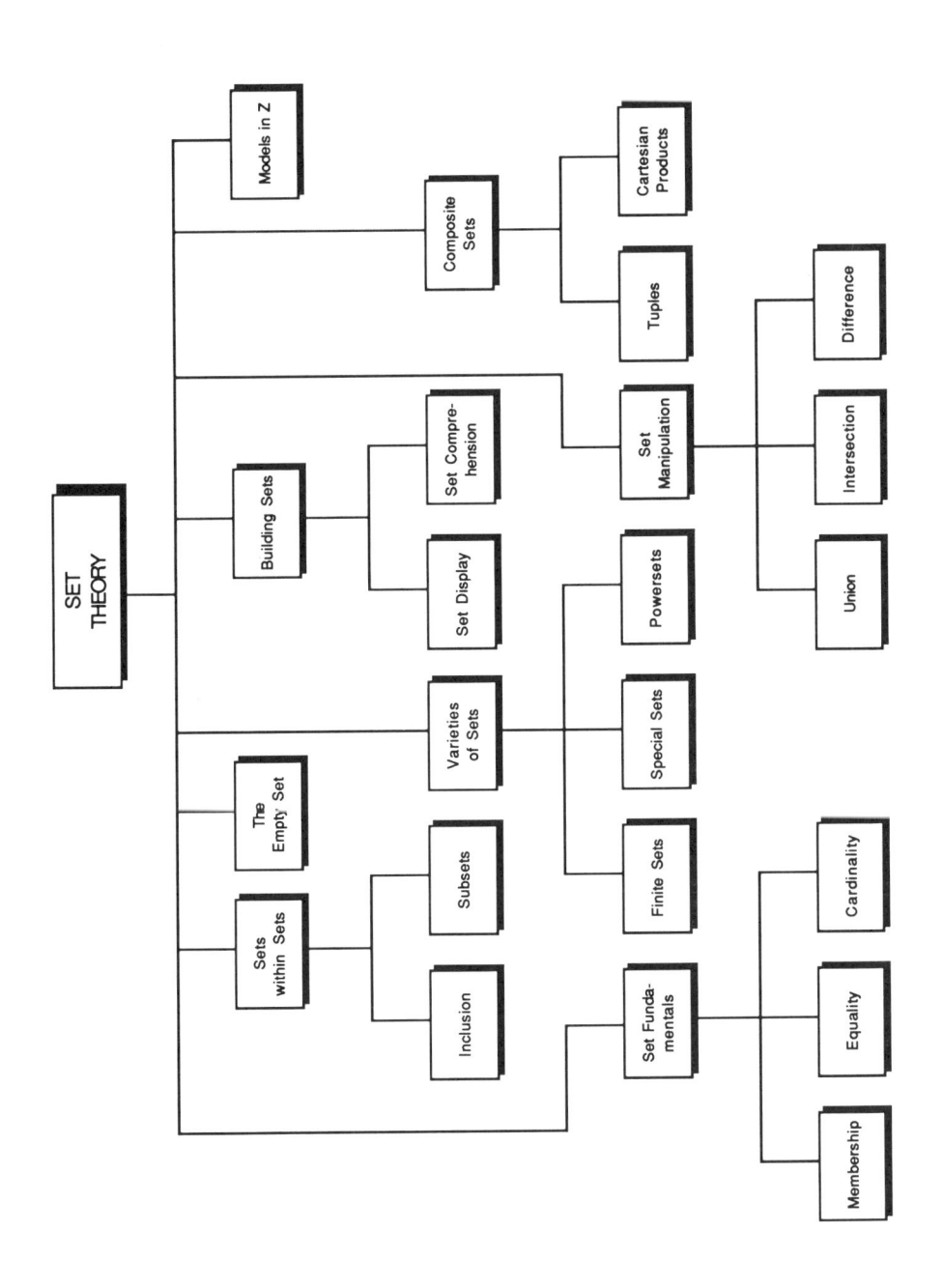

```
SET THEORY
├── Models in Z
├── Composite Sets
│   ├── Cartesian Products
│   └── Tuples
├── Building Sets
│   ├── Set Comprehension
│   └── Set Display
├── Set Manipulation
│   ├── Difference
│   ├── Intersection
│   └── Union
├── Varieties of Sets
│   ├── Powersets
│   ├── Special Sets
│   └── Finite Sets
├── The Empty Set
├── Sets within Sets
│   ├── Subsets
│   └── Inclusion
└── Set Fundamentals
    ├── Cardinality
    ├── Equality
    └── Membership
```

# CHAPTER THREE

# Logic

Logic is extensively used in Z Specifications. On the surface, it provides us with the 'glue' necessary to stick together the building blocks provided by set theory. However, more importantly it provides the specifications with their rigorous and exacting mathematical properties, without which we would not be able to prove anything about the models of the information systems we wish to study.

This chapter introduces the basic concepts behind the logic used in Z Specifications. The treatment of logic will be to a level of rigour appropriate to an introductory text. Our aim is to understand specifications written in Z, and the logic we will cover will be enough to enable us to do that. There is a lot more to logic—we will merely be scratching the surface of what is a very wide area of intense theoretical study. Those readers who wish to explore the mathematical proof of specifications will need to refer to other books which examine the subject in more detail[1].

---

[1] A classic example is *Beginning Logic* by E. J. Lemmon (Van Nostrand Reinhold, 1965). *Logic for Mathematicians* by A. G. Hamilton (Cambridge University Press, 1978) is also worth a look.

# 3.1 PREDICATE LOGIC

For our purposes, a **predicate** is an expression of a relationship between objects which ultimately evaluates to either true or false, not both, and not any other value[2].

Elementary predicates can be formed using $\in$ and $=$. For example, the predicate

$x \in A$

is true if and only if $x$ is a member of the set $A$. This is a fact which is dependent on the object $x$ and the contents of the object set $A$. There were many examples of predicates in Chapter Two.

A predicate is said to be **satisfied** by those values of the objects in the predicate expression which cause it to evaluate to true. So, if $A = \{a, b\}$, then $x = a$ satisfies the predicate $x \in A$.

## Other Predicates

There are other symbols which can be used to form predicates, such as the following relational operators:

| OPERATOR | MEANING |
|---|---|
| $>$ | greater-than |
| $<$ | less-than |
| $\geq$ | greater-than or equal to |
| $\leq$ | less-than or equal to |

However, these symbols are actually abbreviations of $\in$ (a fact which will be explained in Chapter Five).

---

[2] This definition is not strictly correct! Predicate logic is said to be *undecidable*, i.e. there are some expressions in predicate logic for which no technique exists to determine their truth or falsity. In Z we tend to avoid such predicates. More will be said about this later in the chapter.

## Example 3.1

Normal arithmetic expressions, such as "$6 + (y * 8)$", have many possible resulting values. Predicates evaluate to either true or false. Consider the following predicates:

- $x > y$     This predicate is satisfied (is true) precisely when $x$ is greater than $y$. If $y$ is greater than or equal to $x$ then the predicate is false.

- $x \in \{x\}$     This predicate is always satisfied because $x$ is always a member of the singleton set containing $x$.

- $y \in \varnothing$     This predicate is never satisfied because no object can ever be a member of the empty set.

There are two special predicates in the Z notation. They are known as the **logical constants**:

✶ *true*
   ...denotes the predicate which is always true.

✶ *false*
   ...denotes the predicate which is always false.

Thus, if we had an expression depicting a relationship which is always satisfied (such as $x = x$) then we could replace it with *true* and still preserve the meaning of the expression.

## Constraints

In Z Specifications we may wish to enforce certain relationships between objects to say something about our computer model. To do this we *assert* a predicate to be true in the model. For example, we may wish to stipulate that the weight of luggage carried by each passenger on an aeroplane should not exceed 20kg:

$max\_weight \leq 20$

The above predicate is called a **constraint**. It asserts a relationship which must always be preserved from then on in the specification, and hence that the computer system should never be in a scenario where *max_weight* exceeds 20. It is a constraint because it constrains the possible values the variables involved can take on. It works very like the filters used when defining sets such as $\{x : \mathbb{N} \mid x > 10\}$ here the predicate constraint filters out all those $x$'s which are not greater than ten from the resulting set.

## 3.2 OPERATIONS ON PREDICATES

We can form larger, more complex predicates from simpler ones by using logical operators known as **modifiers** and **connectives**. A modifier effects the value of a single predicate, whilst a connective takes two predicate operands and forms a value for the combined predicate based on the individual values of the operands. Combining predicates is said to form a **compound** predicate. The basic building blocks for these are the elementary predicates which are not composed of any modifiers or connectives.

### Equivalence

A very important concept in logic is that of **equivalence**. Two predicates are said to be equivalent if and only if they take the same value as each other.

*      $P \Leftrightarrow Q$
...denotes the predicate "$P$ is equivalent to $Q$".

Note that $P \Leftrightarrow Q$ is itself a predicate. What we have done is to get two predicates and use a logical connective, the equivalence operator, to join them together to form another predicate.

### Example 3.2

A car manufacturer only supplies certain combinations of seating colours and car body colours. The compound predicate

$$car\_colour \in rainbow \Leftrightarrow seat\_colour \in \{black, grey\}$$

would clearly be satisfied if

$car\_colour = $ blue and $seat\_colour = $ grey,

but not when

$car\_colour = $ black and $seat\_colour = $ grey.

Suppose we assert the constraint

$$P \Leftrightarrow Q$$

in a specification. This is stating that in *all* possible legal scenarios that the system may be in, if $P$ is true then so is $Q$, and if $P$ is false then so is $Q$. Such an equivalence means that we can substitute all occurrences of $P$ by $Q$ without affecting the correctness of the specification.

*Discussion*

Equivalence is very useful because it means we can simplify Z Specifications by deriving simpler, but equivalent, expressions and using them instead. This can make specifications clearer. More importantly, it means we can make use of **equivalence transformations**. This is significant because it means we can start with a high-level abstract specification of an information system which we prove is correct and then *refine* the specification by introducing more and more detail until eventually we end up with a low-level (more concrete) model which can more directly be converted into a computer program. The point here is that at each stage of refinement we must show that the lower-level model is *equivalent* to the higher-level model it was derived from. If we can do this then essentially we greatly reduce the possibility of errors creeping in to design and coding.

# Negation

There is one logical operator which is a modifier. When applied to a predicate $P$ it evaluates to the complement (**negation**) of the value of $P$, i.e. if $P$

was false in a given scenario then the resulting predicate would be true, and vice versa.

* $\neg P$
 ...denotes the predicate "not $P$", i.e. the negation of $P$.

The behaviour of the negation operator closely follows that of the English word "not", e.g. "it is *not* raining" is the negation of "it is raining".

### Example 3.3

The following are examples of the negation operator:

- $\neg\text{true} \Leftrightarrow \text{false}$
- $\neg\text{false} \Leftrightarrow \text{true}$
- $(A \neq B) \Leftrightarrow \neg(A = B)$
- $\neg(5 = 5) \Leftrightarrow \text{false}$
- $\neg(x = 2)$ is true precisely when $x$ is not equal to 2

Note that the first four examples are always satisfied, while the final example is only satisfied when $x \neq 2$.

## Disjunction

The result of using the **disjunction** connective operator corresponds closely to the English word "or" (we mean "inclusive-or" here), i.e. when either $P$ or $Q$ are true or both are true, then their disjunction is true.

* $P \vee Q$
 ...denotes the predicate "$P$ or $Q$", i.e. the disjunction of $P$ and $Q$.

There are some equivalences which surface on a closer examination of the behaviour of disjunction; they are known as the laws of or-simplification:

- $P \vee P \Leftrightarrow P$
- $P \vee \text{true} \Leftrightarrow \text{true}$
- $P \vee \text{false} \Leftrightarrow P$

The above three expressions should be read as the LHS being equivalent to the RHS with the ⇔ acting as the separator. For example,

$P \vee P \Leftrightarrow P$

should be read as

$(P \vee P) \Leftrightarrow P$

i.e. we evaluate $P \vee P$, the LHS, first. Later on we will discuss how to interpret such expressions without having to resort to parentheses.

The three expressions are merely stating some obvious facts. Let us consider the first. If we have a predicate $P$ and we form another predicate $P \vee P$ then this predicate is true when $P$ is true and is false when $P$ is false. In other words, we can substitute all occurrences of $P \vee P$ in a specification with $P$ without affecting its correctness.

## Truth Tables

For the disjunction of two predicates $P$ and $Q$ there are two possible values for $P$ (i.e. *true* or *false*) and two possible values for $Q$ thus there are 2*2 possible cases to cater for when combining $P$ and $Q$, and the resulting value of the connected expression depends on the particular combination. This resulting value can be most easily shown in a **truth table** where T represents *true* and F represents *false*:

| $P$ | $Q$ | $P \vee Q$ |
|---|---|---|
| T | T | T |
| T | F | T |
| F | T | T |
| F | F | F |

Each column represents one predicate, each row of a column represents one possible value of the predicate. Let us interpret what this truth table shows. When the predicate $P$ is true and the predicate $Q$ is true then the combined predicate $P \vee Q$ is true. Similarly, when either $P$ or $Q$ are true then $P \vee Q$ is true. Otherwise, $P \vee Q$ is false. While this example is fairly trivial, the graphical nature of the truth table often makes them easier to use and useful in finding the resulting value of more complicated expressions involving several connectives.

The truth table for the negation of a predicate is fairly obvious. It only has two rows since this exhausts the number of different values the predicate to be negated can take on:

| $P$ | $\neg P$ |
|---|---|
| T | F |
| F | T |

Recall that when we assert a predicate in a Z Specification we are saying it is true. Thus if we assert a compound predicate we are only interested in values of the component elementary predicates which are on the same row in its truth table as a 'T' in the final column.

As an example, let us pretend there was a logical connective $\Diamond$ such that the predicate $P \Diamond Q$ had the following truth table:

| $P$ | $Q$ | $P \Diamond Q$ |
|---|---|---|
| T | T | F |
| T | F | T |
| F | T | T |
| F | F | F |

If we asserted $P \Diamond Q$ in a specification we would be saying that one of $P$ and $Q$ is true, no more and no less, i.e. they can never be both true or both false in the computer system being modelled—we are enforcing them as mutually exclusive relationships.

A predicate which has 'T' in every row of its column in a truth table, and hence in every possible scenario, is said to be a **tautology**. Examples of tautologies can be found in Example 3.3.

# Conjunction

The next connective that we will consider is the **conjunction** operator. It is the logical equivalent to the English word "and". If we are given two predicates $P$ and $Q$, then their conjunction is the predicate which is true when $P$ is true and $Q$ is true.

*     $P \wedge Q$
      ...denotes the predicate "$P$ and $Q$", i.e. the conjunction of $P$ and $Q$.

The truth table for the conjunction of two predicates is as follows:

| $P$ | $Q$ | $P \wedge Q$ |
|---|---|---|
| T | T | T |
| T | F | F |
| F | T | F |
| F | F | F |

The above truth table along with the truth table for disjunction suggest the following equivalences known as the laws of and-simplification:

- $P \wedge P \Leftrightarrow P$
- $P \wedge true \Leftrightarrow P$
- $P \wedge false \Leftrightarrow false$
- $P \wedge (P \vee Q) \Leftrightarrow P$
- $P \vee (P \wedge Q) \Leftrightarrow P$

# Implication

The last connective that we will consider is the **implication** operator. It is the one people often have the most difficulty understanding. The main reason for this is that there isn't really an adequate way of expressing the concept in English as succinctly as the previous operators.

We say that "P implies Q" is true except where P is true when Q is false. Informally, what we are saying is that if P is false then we don't care whether Q is true or false—the expression is satisfied whatever the value of Q. However, if P is true then Q must also be true for the statement to be satisfied. It is like saying "if P is true then Q is true" (which of course is a statement which can be true or false).

*     $P \Rightarrow Q$
  ...denotes the predicate "P implies Q" or "if P then Q".

The truth table for implication over two predicates may help clear up any doubts:

| P | Q | $P \Rightarrow Q$ |
|---|---|---|
| T | T | T |
| T | F | F |
| F | T | T |
| F | F | T |

The order of the operands in an implication is significant. $P \Rightarrow Q$ is different to $Q \Rightarrow P$. We say that the predicate before the implies operator is called the **antecedent** and the predicate after the implies operator is called the **consequent**, i.e. antecedent $\Rightarrow$ consequent.

| | | ❶ | ❷ | ❸ | ❹ | |
|---|---|---|---|---|---|---|
| P | Q | $P \Rightarrow Q$ | $Q \Rightarrow P$ | ❶∧❷ | $P \Leftrightarrow Q$ | ❸⇔❹ |
| T | T | T | T | T | T | T |
| T | F | F | T | F | F | T |
| F | T | T | F | F | F | T |
| F | F | T | T | T | T | T |

**Figure 3.1**

## Example 3.4

Figure 3.1 is a truth table which shows an equivalence. It shows the truth values of combining predicates in various ways, and ultimately shows that two of the columns hold exactly the same values in each of their rows. These two columns correspond to two predicates which may be swapped for one another in a specification.

Note that some of the columns are numbered to save having to write the whole expression out in full in the later columns.

The last column is effectively the predicate

$(P \Leftrightarrow Q) \quad \Leftrightarrow \quad ((P \Rightarrow Q) \wedge (Q \Rightarrow P))$.

Since it is true in each row of the truth table, it is a tautology and the equivalence holds. This means we can write $(P \Rightarrow Q) \wedge (Q \Rightarrow P)$ instead of writing $P \Leftrightarrow Q$ in a specification, and vice versa.

## Exercise 3.1

By building a truth table show that the following important expressions are tautologies:

a) $\neg(\neg P) \Leftrightarrow P$

b) $P \vee \neg P \Leftrightarrow true$

c) $P \wedge \neg P \Leftrightarrow false$

d) $\neg(P \wedge Q) \Leftrightarrow (\neg P \vee \neg Q)$

e) $\neg(P \vee Q) \Leftrightarrow (\neg P \wedge \neg Q)$

f) $(P \Rightarrow Q) \Leftrightarrow ((\neg P) \vee Q)$

## Precedence Rules for Logical Expressions

Logical expressions can contain many operators and operands, just like normal arithmetic expressions. In such situations we need rules which tell us how to interpret the expression in order to find its value.

## Example 3.5

Consider the expression

$$A \vee B \wedge C \Rightarrow D \Leftrightarrow E$$

Now, how do we interpret this? When in doubt parentheses should be used to make the order of evaluation explicit:

$$(((A \vee B) \wedge C) \Rightarrow D) \Leftrightarrow E$$

This reveals that $A \vee B$ is evaluated first. The resulting predicate is evaluated as a disjunction with $C$ which in turn leads to a predicate evaluated as the antecedent of an implication with $D$. Finally, the predicate resulting from the implication is evaluated as an equivalence with $E$.

Generally, sequences of the same logical operator are evaluated from left to right; so

$$A \vee B \vee C$$

is equivalent to, and could be written as

$$(A \vee B) \vee C.$$

There is one exception: sequences of implications are evaluated from right to left, i.e.

$$A \Rightarrow B \Rightarrow C$$

is equivalent to

$$A \Rightarrow (B \Rightarrow C).$$

Expressions with non-identical logical operators and without parentheses to help us evaluate them need the use of the following rules called **precedence rules**:

I. Negation has the highest precedence, e.g. "$\neg A \wedge B$" is equivalent to "$(\neg A) \wedge B$".

II. The priority of the remaining operators is, in order of decreasing priority; conjunction, disjunction, implication, and equivalence, i.e.

conjunction has a higher precedence than the other three, and equivalence has the lowest precedence of all.

### Example 3.6

The following are equivalences according to the precedence rules given above:

- "$A \vee B \wedge \neg C \Leftrightarrow D$" is equivalent to "$(A \vee (B \wedge (\neg C))) \Leftrightarrow D$"
- "$A \Rightarrow B \vee C$" is equivalent to "$A \Rightarrow (B \vee C)$"
- "$A \Leftrightarrow B \Rightarrow C$" is equivalent to "$A \Leftrightarrow (B \Rightarrow C)$"

*Discussion*

Z should be all about clarity. Z Specifications should be written with the aim of communicating the model as swiftly as possible and not leaving it open to misinterpretation. This means that making certain things explicit can be very helpful. If there is a logical expression (or any other form of expression) then enforcing the above rules with the use of parentheses, although not essential, is a good idea as it leaves little room for error on behalf of the reader.

## 3.3 QUANTIFICATION

There are three more important notational constructs used in Z Specifications based around predicate logic. They are powerful because they offer a short notational abbreviation for potentially infinite length logical expressions.

The general name given to the idea behind these constructs is **quantification**. You may already be familiar with the quantification in ordinary mathematical arithmetic. For example, suppose we wish to add up the squares of all the numbers between one and a hundred, i.e. we wish to find the value of the expression

$$1^2 + 2^2 + 3^2 + 4^2 + 5^2 + \ldots + 98^2 + 99^2 + 100^2$$

It is very tedious to have to write the whole sequence out in full. We 'cheated' by using an ellipses to mean "you can guess what goes here so we won't bother writing it!". Mathematicians use a piece of notation called the **summation quantifier** (a capital Greek sigma symbol) to express such structures, i.e. sequences of additions with some identifiable pattern. The above example can be represented with the shorthand

$$\sum_{i=1}^{100} i^2$$

which denotes "1 + 4 + 9 + 16 +...+ 9604 + 10000". Thus, the above notation represents a series of **terms** which are added together over a given **range**. A *term* is one of the $i^2$ values, and the *range* is the set of values that $i$ can take, i.e. from 1 to 100 inclusive.

Notice that the $i$ used in the quantifier doesn't have any predefined meaning—it is simply introduced as a sort of 'placemarker'. Its name can be changed without affecting the answer. We could have written

$$\sum_{z=1}^{100} z^2$$

to get exactly the same meaning, and hence exactly the same answer. Thus, it is not a real variable (or object) and is called a **bound variable**—it is *bound* to the quantifier.

When we use quantification in Z, the range is specified using a set, and a bound variable is used as above. However, we must be careful not to give a bound variable a name used by another object which already exists in a specification otherwise confusion will arise.

## Existential Quantification

**Existential quantification** is the name given to a logical form of quantification which represents the disjunction of a series of predicate terms over a given range. The predicate formed by existential quantification is satisfied when one or more of the disjuncts are true.

For example, suppose we have the predicate

$x \in B$

where $x$ is a bound variable (like $i$ in the summation quantifier), and we wish to disjoin this predicate ranging over the set $A$, then if $A = \{1, 2, 3\}$ our existential quantification will be equivalent to

$((1 \in B) \vee (2 \in B) \vee (3 \in B))$

which forms a predicate which is true when any of the disjuncts are true (see the truth tables for disjunction in the previous section to convince yourself of this).

**✱** $(\exists D \bullet P)$
...denotes the predicate which is true precisely when there exists at least one D that satisfies $P$.

The D above denotes a **declaration** which introduces the bound variables. We have already come across a Z declaration. In Chapter Two when we used implicit notation for set construction we had a declaration of $x : T$. This introduced a new object called $x$ which was of type $T$. In general, a declaration can introduce several new objects (all of which must have a specified type), and each of these objects must be separated by a semi-colon. For example the declaration

$x : APPLE; y : PEAR$

introduces two new objects, one taken from the set $APPLE$ and the other from the set $PEAR$. We will have much more to say about declarations in the next chapter.

We can express the previous example of existential quantification in Z as

$(\exists x : A \bullet x \in B)$.

It is a predicate which, when asserted, is saying "there exists an $x$ in $A$ such that $x$ is a member of $B$". It is saying that there is at least one of the members of set $A$ in set $B$.

Sometimes we may wish to restrict the range of quantification, e.g. we may choose not to examine all members of set $A$ but just the odd numbers in $A$. We do this by **'filtering'** the set in the declaration using an additional predicate (in much the same way as we created sets implicitly in Chapter Two by filtering a larger set with a predicate to form a subset which we then named).

\*    $(\exists D \mid Q \bullet P)$
      ...denotes $(\exists D \bullet Q \wedge P)$.

Here we are not just checking for the existence of a D which satisfies $P$, but for a D which satisfies $P$ *and* satisfies $Q$. In the above example, if we just wanted to consider all the odd numbers in the set $A$ then we would have introduced the filtering predicate "$x \bmod 2 \neq 0$" (which is true precisely when $x$ is odd) to give us the predicate

$(\exists x : A \mid x \bmod 2 \neq 0 \bullet x \in B)$.

### Example 3.7

Suppose we have the predicate

$(\exists x : \mathbb{N} \mid x > 10 \wedge x < 100 \bullet x \in A)$.

This states that there exists a nonnegative whole number (a natural number), greater than 10 and less than 100, in set $A$.

The declaration specifies that the range of quantification is over all the positive numbers, but the predicate "$x > 10 \wedge x < 100$" reduces the range so that only numbers between 10 and 100 (exclusive) are considered.

## Unique Quantification

Existential Quantification forms a predicate which states that *at least one* of the predicate terms is true. However, sometimes it may be useful to state that *exactly one* term should be true—no more, no less.

* $(\exists_1 D \bullet P)$
...denotes the unique existence of a D that satisfies $P$.

## Example 3.8

Suppose we had a multi-user computer system with many terminals from which users could access a minicomputer. Each user has their own password to enter the system. Users are granted varying levels of power. Level 5 users have the most power—they have access to all the files and can generally control the system.

Let us model this system by having a set *users* which contains ordered pairs of each users' password and the level of that user. We want to state that there can only ever be one user who is at level 5. The following predicate states this:

$(\exists_1 x: PASSWORD \bullet (x, 5) \in users)$

## Universal Quantification

The last form of quantification we will cover specifies a predicate which is satisfied only when all its predicate terms are true. **Universal quantification** represents the conjunction of a series of predicate terms over a given range.

For example, in a similar way to the example of existential quantification, let us suppose we have the predicate

$x \in B$

and we wish to conjoin this predicate ranging over $A$, then if $A = \{1, 2, 3\}$ our universal quantification will be equivalent to

$((1 \in B) \wedge (2 \in B) \wedge (3 \in B))$

which forms a predicate which is satisfied when all of the conjuncts are satisfied.

* $(\forall D \bullet P)$
  ...denotes the predicate which is true when $P$ is satisfied for every D.

A closer look at the previous example shows that it is equivalent to the subset operator on $A$ with respect to $B$:

$$A \subseteq B \Leftrightarrow (\forall x : A \bullet x \in B)$$

The RHS of the above equivalence is saying "for each member of $A$ (call it $x$), this $x$ is a member of $B$". Again, note that we can call each member anything we want—it is just a bound variable—but we must ensure that the name we used cannot get confused with other names in our specification.

As in existential quantification, we may wish to restrict the range of quantification, e.g. we may chose not to examine all members of set $A$ in the above example but just those numbers in $A$ greater than one. Again, we do this by filtering the set in the declaration using an additional predicate.

* $(\forall D \mid Q \bullet P)$
  ...denotes $(\forall D \bullet Q \Rightarrow P)$.

Here we are not checking that every object in D satisfies $P$, but just those D which satisfy $Q$. Those objects that do satisfy $Q$ must then *also* satisfy $P$. That is why we use the implication operator—the predicate $Q \Rightarrow P$ is satisfied when $Q$ is false or when *both* $Q$ and $P$ are true. It is ensuring that the quantification is only not satisfied if a $P$ is false when its corresponding $Q$ is true. We don't care if a $P$ is false if its $Q$ is false because we are only interested in the cases when $Q$ is true. We filter out unwanted conditions.

In the above example, if we just wanted to consider all those numbers in the set $A$ greater than one then we would have introduced the filtering predicate "$x > 1$" to give us the predicate

$$(\forall x : A \mid x > 1 \bullet x \in B).$$

## Example 3.9

If we have a set of natural numbers, we can always pin-point one of its members as the **least member**, i.e. it is less than all the other members of the set. Suppose $lm$ is the least member of set $A$, we can create the following predicate to assert this fact:

$(\forall y:A \bullet lm \leq y)$

Since $lm \leq lm$ is a tautology, the above is satisfied only when $lm$ is the least member of $A$. It reads "for every $y$ in $A$, $lm$ is less than or equal to this $y$".

There is a strong link between universal quantification and existential quantification. In fact, we can define one in terms of the other. The following two equivalences show the link:

■  $(\forall D \bullet P) \Leftrightarrow \neg(\exists D \bullet \neg P)$

which is saying "for every D, P is true" $\Leftrightarrow$ "there does not exist a D for which P is false". If something is true in every case then there does not exist a case when it is false.

■  $(\forall D \bullet \neg P) \Leftrightarrow \neg(\exists D \bullet P)$

i.e. "for every D, P is false" $\Leftrightarrow$ "there does not exist a D for which P is true". If something is false in every case then there does not exist a case when it is true.

*Discussion*

Quantification in logic is a notational shorthand for potentially infinite sequences of predicate terms. Consider a modified least member predicate:

$(\forall y:\mathbb{N} \bullet x \leq y)$

This could be re-written as the conjunction of an infinite number of predicates (each one having a different natural number to replace $y$). Thus, by considering each predicate term in turn it could take an infinitely long time to decide whether the quantified predicate expression were true or false! This is why predicate logic is said to be **undecidable**.

From this it should be clear that, in the general case, truth tables cannot be used to find out the truth or falsity of a predicate. This is because we would require an infinite number of columns for quantification over an infinite set. Truth tables are useful for showing the properties of the logical operators, and for evaluating expressions involving a manageably small number of different predicate terms. Chapter Nine introduces the normal way of evaluating generalised predicates, but such techniques are quite complicated and need not be studied if just an introduction to Z is desired.

**Example 3.10**

If we have a finite set $A$ of natural numbers (i.e. $A$ is of type $\mathbb{PN}$) and an operation called *max* which finds the maximum of a set (the largest number in the set), then *max A* is the maximum of set $A$. The following predicate assert the properties of *max A*:

$$(A \neq \emptyset) \wedge (\forall x: \mathbb{N} \mid x \in A \bullet x \leq max\,A) \wedge (max\,A \in A)$$

The above predicate asserts that set $A$ must not be empty (otherwise it wouldn't have a maximum), that "for every natural number which is a member of $A$, that number is less than or equal to $max\,A$", and that "$max\,A$ is a member of $A$". These last two conditions ensure that $max\,A$ is indeed the largest number in set $A$. Again, $x \leq x$ is always true (it is a tautology).

We could have written the predicate as

$$(A \neq \emptyset) \wedge (\forall x : A \bullet x \leq max\,A) \wedge (max\,A \in A)$$

which is equivalent to the former version.

# 3.4 SUMMARY

A predicate is an expression which may evaluate true or false. The predicate

*my_name* = fred

can be true or false. In a specification we sometimes want to say categorically that something is true, so if we asserted the above predicate we know that the variable *my_name* would always be equal to fred. Similarly, if we asserted

$$\neg(\mathit{my\_name} = \mathtt{fred})$$

then we would be saying that the variable *my_name* could never be equal to fred.

In this chapter we have introduced some very important notation which correspond to important fundamentals of Z. Logic provides the backbone of Z Specifications. Without it they would be of little use. It allows us to take simple statements of fact and build more complex ones using the logical connectives. More importantly, it provides the computer technician with a framework for rigorously proving that a specification is consistent via logical inference. We now have the apparatus to unambiguously state some complex requirement in a short space in an understandable way.

# CHAPTER FOUR

# Building Z Specifications

The previous two chapters described the mathematical foundations of Z Specifications. From these working materials the Z methodology provides the techniques needed to model information systems. This chapter describes the format of Z Specifications. It shows how set theory and logic may be used to form specification documentation; it shows how to interpret the notation and conventions used in Z, and develops simple realistic examples to illustrate each new concept as they are introduced.

It is very important to try to understand everything in this chapter since every Z Specification is based on the material within it. It describes the 'skeleton' of a specification document. However, it will not be until the next few chapters, which cover some important mathematical tools corresponding to the 'meat' of such a document, that you will be able to understand the Z Specifications written by others.

# 4.1 OBJECTS IN Z

Up until now we have used the term 'object' rather loosely. Let us recap what we know about objects in Z:

- The basic idea is that an object is an abstract representation (in the specification) of some concrete entity (in the real world). Each object is associated with some identifiable thing, be it real or imaginary, which is useful to the specification. That object *represents* something we wish to model.

- In computer programming there are **variables**. These are names which represent stores in memory that can hold different values that vary during execution of a program. Z has variables which may represent different real-world objects over the life of the system.

- An identifier is a name given to an object in Z which consists of a sequence of characters. These characters may be upper or lower case letters of the alphabet, digits, or underscores (i.e. "_"). Identifiers must start with a letter.

### Example 4.1

The following are examples of identifiers in Z. Note the convention in this book that all identifiers appear in a distinct typeface:

- *OutOfStock* – this is the name of the operation that deletes the code of items no longer in stock in the Shopping Catalogue example of Chapter Two.
- {red, amber, green} – this is an object (a set in this case) which is composed of three named objects (fixed objects). We could associate this set with the identifier *trafficlight*. This name now serves to identify the set so we don't have to write out the set in full each time we wish to refer to it.

Identifiers are very important in Z. Being able to give a name to some concept or entity is a powerful capability. We can take a large or complicated structure (the components of which we may or may not understand) and give it a name. From then on, when we wish to think of the structure or refer to it in our specifications, we simply think or write its name knowing that the name alone represents the structure we must deal with. Essentially, we are using abstraction—working with names as much as possible means we can abstract away from the details which the names 'hide' and work at a higher level. It means we can more easily deal with very large and complicated systems.

## Declarations

Chapters Two and Three have shown that variables in Z consist of an identifier and its corresponding type. When we introduce a new identifier in a Z specification we must state its name and the set from which it may draw its values. For example, in programming we may introduce a variable called $x$. There is usually a limit to the greatest value this variable can take. Let us say that the programming language can only store numerical values between -32768 and 32767 inclusive. In this case the value of $x$ can only be taken from the set

    -32768..32767

(using the range notation introduced in Chapter Two).

When we introduce an identifier we must **declare** it. This involves writing the name of the identifier followed by a colon and a set which corresponds to the type of the identifier.

*     $x : A$
    ...denotes the declaration of $x$ which is of type $A$.

*     $x, y : A$
    ...denotes the declaration of several different identifiers, separated by commas, all of which are of type $A$.

Notice that you can easily introduce several identifiers which are all of the same type—both $x$ and $y$ are introduced and are of type $A$. There may be some situations where we want to declare several different identifiers of different types on the same line:

*     $x : A;\ y : B;\ w, v : T$
  ...denotes three separate declarations on one line, each of which are separated by a semicolon.

In general it is not a good idea to cram too much information on one line; the emphasis should always be on clarity for the reader.

## 4.2 Z SCHEMAS

An information system, by definition, is something which contains some sort of meaningful data and acts upon this data to produce some (hopefully) useful output. For example, an accounting system may deal with numbers which represent a firm's capital, assets, and liabilities. Thus, the data are the transactions entered into by the firm in the form of account postings consisting of the date a transaction took place, a description of the transaction, and the sum of money involved. In this system the useful output may consist of such functions as the production of correctly formatted balance sheets, credit control facilities, management audit reports, and the general storage of the various ledgers and accounts of a double-entry bookkeeping system.

Z is said to be a *model-based* specification language because it is primarily concerned with modelling the *data* involved in the information system to be specified, and then modelling the operations which transform the data into some useful form. In Z, we do this using various mathematical entities. We tend to group together sections of these entities into constructs called **schemas**. Informally, a Z schema is a collection of related components of a Z Specification which introduces objects and specifies the relationships between these objects.

A Z schema is presented graphically to highlight its contents within a specification. It normally has two areas: a **signature** and a **predicate part**.

```
┌─── schema name ─────────────────────────────┐
│ signature part                              │
│ ─────────────────                           │
│ predicate part (optional)                   │
└─────────────────────────────────────────────┘
```

The signature contains a declaration of the variables to be used in the schema. The predicate part can show relationships between the variables (sometimes referred to as **axioms** of the variables) declared in the signature. The signature part is above the middle dividing line. The predicate part may be omitted from a schema in which case it will not have a middle dividing line.

✱   The following denotes a schema named $S$ which introduces a variable $x$ of type $A$ that can only take on values satisfying the predicate $P$.

```
┌─── S ───────────────────────────────────────┐
│ x : A                                       │
│ ─────                                       │
│ P                                           │
└─────────────────────────────────────────────┘
```

The box-like property of Z schemas help make them stand out in a specification document. This is good because all the important mathematical concepts can be easily found in large documents.

Z schemas are modular in nature—each introduce part of the entire specification, a little at a time, in named blocks. Naming schemas brings in the benefits of abstraction:

■   We don't have to remember the precise specification of the operation. For example, to add a new item of stock to the inventory we need only remember that the schema *AddNewStock* provides such a specification.

■   We are able to treat schema entities as any other object and perform operations on them. For example, we can refer to schemas within other schemas, we can create new schemas from old ones (schema reuse is a useful way of saving space and helping consistency), and we can even have *schema types*—for objects that inherit the properties of a given schema.

The modularity and graphical clarity of Z schemas is probably the main reason why Z is gaining popularity relative to other formal methods.

Another important idea exploited by Z schemas is that relationships between the variables involved in the specification can be explicitly shown. Imagine a specification which had many objects. Each could appear to represent seemingly unrelated components of the system. Such a specification would be hard to follow. If instead the relationships between the objects were explicitly documented, it would be easier to build up a mental picture of what was going on and how everything worked together in the scheme of things. The reader would thus gain a deeper understanding of the system. The multitude of objects would then not seem so different after all.

## Example 4.2

Suppose there was a model of a check-in service for an airline. There would be many different entities involved and thus many identifiers. We could have the following identifiers in the specification:

*checked* : $PCODE$

*luggage, maxweight, seats, smoking* : $\mathbb{N}$

Of course, the Z specification would have to say what each of these names meant—stating their type is not enough.

We will now explain what each identifier represents and state any relationships between them:

■   *checked* is the set containing the flight ticket code numbers of the passengers who have checked in.
■   *seats* is the number of available seats on the flight.

Obviously, the number of passengers that check-in must not exceed the number of seats on the flight, therefore the following constraint can be made:

$\#checked \leq seats$

This predicate is asserting that the number of passengers checked-in (which can be found by counting the number of members of the set *checked*) must always be less than or equal to the number of seats on the flight.

- *luggage* represents the total weight of the luggage checked-in so far, rounded up to the nearest kilogramme.

If *maxweight* is the maximum weight allowed for passengers and their luggage, and we assume that each passenger weighs on average 75kg, then the following is a constraint on the specification:

(#*checked* * 75) + *luggage* ≤ *maxweight*

i.e. The number of passengers multiplied by their average weight (which gives the total weight of all the passengers) plus the weight of their luggage must not exceed the maximum permissible weight.

*Note*

Firstly, when we introduce numbers in a specification like 75 above then we should say what they represent. A better idea would be to give the number a name so that its meaning was more obvious, e.g. *av_pass_weight* being a reasonable name for average passenger weight. Secondly, we put parentheses around the multiplication to emphasise that it should be calculated first. However, this is not strictly necessary because multiplication (and division) have a higher precedence than addition (and subtraction).

The last constraint which can be stated is the following predicate:

#*checked* * .50 ≥ *smoking*

This asserts that no more than half (50%) of the number of passengers should be allocated seats in the smoking area. Again, .50 could be named for greater clarity.

The above is a simple example of stating relationships in specifications. In more complex models the relationships might be less obvious to the reader who wants to learn about the system, and making them explicit will be even more important.

A Z schema shows the relationships between the objects it introduces by constraining the values each of them can take to those which make sense in the specification. This is done with a predicate in the predicate part. When the predicate part is omitted the value of the predicate is taken as equivalent to *true* by default, which means there is no constraint on the values the objects in the signature can take.

## State Space

The first thing we must do when specifying a system is build a model of the data involved. What we want to do is model the data in terms of set theory and show the important relationships between the components of the system. The model must cover all the possible scenarios which the information system may legally be in. For example, we shouldn't use a simple set to model something which may feasibly have a repeated element in it, e.g. a set modelling a list of friends invited to a party doesn't distinguish between two people with the same forename and thus cannot successfully model the party list in all possible scenarios. Something other than a simple set must be used.

To model the data we introduce various objects into the specification. Each object represents some aspect of the system. For example, to specify a "desktop diary" program which keeps an appointment book we may use a set *appointments* of 3-tuples which hold the time of an appointment, its duration, and details of who the appointment is with. This models the real world using mathematical entities.

*Note*

So far in this book we have covered only basic set theory. This is not really enough to build readable specifications of most systems because real-world data is usually far more complex. Later chapters will show simple, yet powerful tools built on set theory which will help us build more elegant system models. Until then we will be limited to fairly simple examples.

The relationships between the mathematical objects we introduce to model data will hold throughout the specification. These relationships are *invariant*—they never change.

So, all Z Specifications must start with a definition of the objects used in the model and the invariant relationships between them which constrain their possible values. This definition describes all the possible legal states any one of which the system may be in at any one moment in time. A scenario which doesn't fit in the model is an illegal scenario which the system should never be in if it is working correctly. In Z, the set of all legal states of a model is called the **state space** of the model.

The concept of state space is very important. It means that we can take a tight control over the system, and we can thus make sure that, whatever values the variables take, we know that they make sense in so far as they represent a stage in the system from which a correct result can follow.

We can display the state space of a model using a Z schema. This schema is said to denote the **static** elements of a specification because its contents assert aspects of the model which are invariant and do not change the state of the model—they define the state space but do not mention how the system may move from one state to another.

### Example 4.3   HEP CONTROL SYSTEM

Suppose we have a simple model of a control system for a hydroelectric powerstation (HEP) which consists of an upper reservoir fed by water from the mountains, a large concrete dam, and a series of tunnels through the dam which contain turbines that convert the energy of fast-flowing water into electricity. The level of the upper reservoir is read-off by an electronic device linked to the main computer. Readings are in centimetres rounded up to the nearest whole number. The control system works by regulating the level of the water in the reservoir. The state space of the system is given by the schema

```
┌──── Reservoir ──────────────────────────────┐
│ level, max, min : N                         │
│─────────────────────                        │
│ level ≥ min ∧                               │
│ level ≤ max                                 │
└─────────────────────────────────────────────┘
```

This schema states the following:

- The name of the state space of the system is *Reservoir*.
- The system is modelled by three objects. These objects have values which are members of the set of natural numbers.
- The predicate *level* ≥ *min* must always hold in the system.
- The predicate *level* ≤ *max* must always hold in the system.

Now, given these facts, we can see that the state space of the system does not include the scenario where, say, *level* = 3240 and *max* = 3000. This is because the invariant *level* ≤ *max* would not be true if that were the case.

The state space contains all those states (levels of the reservoir) which ensure the system is as specified (i.e. the reservoir isn't too high or low). If the specification were incorrect, i.e. it was possible for it to be in a state other than in the state space, then it would be possible to show this in a mathematical proof. This is one of the tremendous advantages of specifications that are based around state space.

## Example 4.4    TELEPHONE DIRECTORY

A telephone directory is modelled as a set *teldir* of ordered pairs. Each pair is of the form (name, number), i.e. the first element of the pair is the name of a person taken from the set *NAME* which contains all possible names, and the second element of the pair is the phone number of the named person taken from the set *NUM* of all possible legal telephone numbers. Thus the type of *teldir* is **P**(*NAME* × *NUM*).

A further variable called *dirsize* holds the number of entries in the telephone directory. The following schema describes the state space of the system:

```
┌─── Directory ────────────────────────────
│ teldir : P(NAME × NUM)
│ dirsize : N
├──────────────────────────────────────────
│ dirsize = #teldir
└──────────────────────────────────────────
```

Notice that we don't really need *dirsize* in the specification—if we wanted to find the number of entries in the directory we could simply refer to the cardinality of *teldir*.

However, variables like *dirsize* improve the readability of Z specifications especially if they represent something quite complex. Instead of having to work out what the author is trying to show, an appropriately named identifier can make it clearer and save a lot of time.

# Initial State

A model must specify how a system starts initially. The state that a computer system is in when it starts is called an **initial state** of the system. Every model must have at least one initial state, but may contain many.

For example, if we were developing software for a new company about to start trading for the first time, then the day the system went 'live' there would be no customer records, stock, etc... The initial state here would be the state where the data sets contained no objects. However, if the software was to supersede existing software for an established firm, then the existing data would be converted over to the new system and the initial state of this new system might be one of many legal states.

Again, we can specify the initial state of a system using a Z schema. This schema must have the same signature and predicate part as the state space schema. The predicate part must additionally specify values for the objects in the signature and these values must be legal so that the set of initial states is a subset of the state space.

## Example 4.5   HEP CONTROL SYSTEM continued...

If we continue with our HEP example then the following schema specifies the set of possible initial states of the model:

┌─ *InitReservoir* ─────────────────────────────
│ *level, max, min* : $\mathbb{N}$
│ ─────────────────────────────────
│ *level* < 1800 ∧
│ *max* = 3000 ∧
│ *min* = 70 ∧
│ *level* ≥ *min* ∧
│ *level* ≤ *max*
└──────────────────────────────────

In this particular example the set of initial states is not very relevant. This is because we don't really care what level the reservoir is initially as long as it is below the maximum and above the minimum! However, let us say that the surveyors have stipulated that initially the water level must be less than 18m so as not to put too much pressure on the dam while they take readings to test its safety.

Notice a general convention that the schema describing the set of initial states is named after the state space with the word "Init" used as a prefix.

## Schema Reference

Example 4.5 specified the initial state of the HEP system by copying out the entries of the state space into its signature and predicate parts as appropriate. However, Z schemas are entities in their own right, and if they are named we can refer to them by just having to write their name. This technique is known as **schema reference**.

*Notation*

When we see the name of one schema, call it $A$, in the signature of another schema (named $B$) then we take it as meaning that the signature of $A$ is added to the signature of $B$, and the predicate part of $A$ is joined to the predicate part of $B$ via conjunction.

## Example 4.6  TELEPHONE DIRECTORY continued...

Let us continue with the telephone directory example. The specification states that the directory starts by being empty. We can show this initial state with the following schema:

$$
\begin{array}{|l}
\hline
\ InitDirectory \\
\hline
\ Directory \\
\hline
\ teldir = \emptyset \\
\hline
\end{array}
$$

The above schema uses schema reference to introduce the variables *teldir* and *dirsize*, and the predicate part of *Directory* which states that *dirsize* is equal to #*teldir*. Thus, we could have written *InitDirectory* in the following fully-expanded form which is equivalent to the above but without making use of schema reference:

$$
\begin{array}{|l}
\hline
\ InitDirectory \\
\hline
\ teldir : \mathbb{P}(NAME \times NUM) \\
\ dirsize : \mathbb{N} \\
\hline
\ teldir = \emptyset \land \\
\ dirsize = \#teldir \\
\hline
\end{array}
$$

Although the two schemas are equivalent, using schema reference is preferable because it is succinct and it is more likely to be consistent—there is less chance of missing out a declaration or part of the predicate which may lead to an incorrect specification which confuses the reader.

Notice that we have said that initially *teldir* is empty, but we have said nothing about the initial value for *dirsize*. In fact, we have done this implicitly because the predicate

$dirsize = \#teldir$

in *Directory* (which is referred to in *InitDirectory*) ensures that the only possible legal initial states are precisely when $dirsize = 0$.

### Exercise 4.1

Specify the initial state for HEP system of Example 4.5 in a schema which uses schema reference.

## Changes of State

An operation generally takes one state of the system (i.e. of the data in the system) and transforms it into another state of the system. A change of state corresponds to changes in the data. Therefore, operations which transform data are **dynamic**—they change state. For example, in the Shopping Catalogue example in Section 2.3 a state of the system was the contents of the sets *instock* and *onorder* at a particular point in time. When some new stock arrived we specified (informally) an operation which changed the contents of the two sets. The state of the system was changed by the operation.

Chapter Two introduced the convention used in Z to distinguish between the state of the system before an operation (the **before state**) and the state after an operation (the **after state**) by appending a dash ( ' ) to the identifier when referring to the after state. For example, the level of the reservoir in the HEP control system before an operation is represented by *level*, and the level after an operation is represented by *level'*.

A dashed schema can be written out in full as having all its component variables (i.e. the variables declared in its signature) dashed.

### Example 4.7   HEP CONTROL SYSTEM continued...

If the state space schema of the HEP system is dashed its fully-expanded form would appear as

―― *Reservoir'* ―――――――――――――――――
$level'$, $max'$, $min'$ : $\mathbb{N}$
―――――――――――――
$level' \geq min' \land$
$level' \leq max'$
―――――――――――――――――――――

Notice that every occurrence of the variables in the signature are replaced by their dashed forms in the predicate part as well.

## Input and Output

Operations on a system tend to involve some form of input or output. For example, the computer in our HEP system receives input from the electronic measuring equipment which monitor the height of the reservoir. In a database query system, the input would be the query entered by the user, and the output would be the computer's response to the query, e.g. a list of employees who have worked for a firm for more than ten years.

*Notation*

The convention in Z for input and output is to suffix a question mark "?" to identifiers which hold input, and an exclamation mark "!" to identifiers which hold output.

### Example 4.8

Consider the following declarations in the signature of a schema in a security system:

$$id? : ID$$
$$status! : \{\texttt{cleared}, \texttt{blocked}\}$$

The value of the variable *id?* is the identification number input to the system by someone meaning to get past a security barrier. The value of the variable *status!* is assigned by the operation depending on the value of *id?*. If *id?* is correct, then *status!* will equal `cleared` otherwise it will equal `blocked`.

To model an operation on a system we must specify its **preconditions** and **postconditions**. The postcondition of an operation is a predicate which describes the set of states the system can be in after a successful and correct attempt at performing the operation is carried out. The precondition is a predicate which describes the set of states the system can be in in order to successfully complete the operation and hence satisfy the postcondition.

Thinking about an operation in terms of its pre- and post-conditions is a useful way of focusing on exactly what the operation should achieve. The postcondition states the goal: what we are trying to achieve with the operation. The precondition then identifies what states we need to be in to achieve this goal. Not only do we have a better chance of specifying all possible scenarios, but we also have a convenient vehicle for specifying *what* an operation should achieve without having to say *how* it is achieved.

### Example 4.9    TELEPHONE DIRECTORY continued...

The process of adding a new name and phone number to the telephone directory is an operation. The new name and new number are inputs to the system which are added to the directory. The operation is specified in a schema called *NewNumber*.

---
*NewNumber*

$Directory$
$Directory'$
$name?: NAME$
$number?: NUM$

---
$teldir' = teldir \cup \{(name?, number?)\}$

---

Notice that there are no preconditions to this operation. It does not matter what name and number is input. The types of the variables *name?* and *number?* ensure that the information added to the telephone directory is meaningful (however, there is nothing stopping the user entering a wrong phone number or a misspelt name—that is something the user must get right!).

The postcondition specifies that the state of the system changes—*teldir* changes and so *dirsize* changes. *teldir* now contains all the ordered pairs it had before the operation plus the new pair.

Note that the set union operator is used. This means that if the new pair already happens to be in the directory then the directory remains effectively unchanged (i.e. essentially *teldir'* remains equal to *teldir* and *dirsize'* remains equal to *dirsize*). Also, we had to put the new pair in curly braces in order to create a singleton set because the set union operation is only defined on sets.

### Example 4.10  HEP CONTROL SYSTEM continued...

As a safety precaution, the engineers at the HEP control centre like to check the level of the reservoir every half hour. They want a reading of its height in centimetres, and a Reservoir Status Message (RSM). The RSM is one of the following messages:

Too_Low         if the reservoir level comes within one metre of its minimum level.

Satisfactory    if the reservoir level is well within safe bounds.

Too_High        if the reservoir level comes within 2½ metres of its maximum level.

The following schema specifies the reservoir check operation:

―――― *CheckRes* ――――――――――――――――――――
*Reservoir*
*Reservoir'*
*current_level!* : $\mathbb{N}$
*RSM!* : {Too_Low, Satisfactory, Too_High}
――――――――――――――――――――――――――
$(level > max - 250 \Rightarrow RSM!$ = Too_High$) \wedge$
$(level < min + 100 \Rightarrow RSM!$ = Too_Low$) \wedge$
$(level \leq max - 250 \wedge level \geq min + 100 \Rightarrow RSM!$ = Satisfactory$) \wedge$
*current_level!* = *level* $\wedge$
*level'* = *level* $\wedge$
*max'* = *max* $\wedge$
*min'* = *min*
――――――――――――――――――――――――――

The above schema is the largest we have encountered so far. Later on we will learn how its size can be slightly reduced.

*CheckRes* introduces two variables other than the usual before and after *Reservoir* state variables. *RSM!* is an output variable which is of type {Too_Low, Satisfactory, Too_High}, i.e. it can take on any one of the three values corresponding to three messages. *current_level!* is an output variable which returns the current value of *level*, i.e. the level of the reservoir.

The predicate part of *CheckRes* contains three implications. The antecedent of each of the implications can be thought of as part of the precondition of the schema. This is because depending on which of the three are true, the value of *RSM!* is decided. For example, on the first line we have:

$(level > max - 250 \Rightarrow RSM! = \text{Too\_High})$

This means, *if* the level of the reservoir is more than the maximum minus 250cm *then* the value of *RSM!* is Too_High.

The final lines of the predicate part assert that none of the state variables change value in the operation, thus there is never a change of state of the system by merely checking the level of the reservoir.

## Δ and Ξ Schemas

We have seen how some schemas (such as *AddNumber* in the telephone directory example) may change the state of a system, whereas others (such as *CheckRes* in the HEP example) never change the state of a system.

Changes in state are denoted by dashed variables. However, to save having to introduce the state space dashed variables for each operation, the Z Notation has a couple of useful conventions.

*Notation*

- A capital delta "Δ" is prefixed to a schema name to introduce a set of before *and* after states for the schema.

- A capital xi "Ξ" is prefixed to a schema name to introduce a set of before and after states such that the after states are equal to the before states.

Hence, for a model with state space $S$, we can refer to $\Delta S$ when an operation may change the state of the system, and refer to $\Xi S$ when an operation never changes the state of the system.

## Example 4.11  HEP SYSTEM continued...

The following schema is a definition for $\Delta Reservoir$:

$$
\begin{array}{|l}
\hline \quad \Delta Reservoir \\
\hline
Reservoir \\
Reservoir' \\
\hline
\end{array}
$$

This is our first example of a schema with no predicate part. We will use it to help our definition of the next schema:

$$
\begin{array}{|l}
\hline \quad \Xi Reservoir \\
\hline
\Delta Reservoir \\
\hline
level' = level \land \\
max' = max \land \\
min' = min \\
\hline
\end{array}
$$

The schema shows that the xi prefix is used in operations whose after states are equal to their before states. The reference to $\Delta Reservoir$ introduces the dashed and undashed variables.

The above is a good example of the power of using schemas. We define $\Delta Reservoir$ in terms of two other schemas, $Reservoir$ and $Reservoir'$, then we can define another schema $\Xi Reservoir$ in terms of $\Delta Reservoir$.

Thus we can create more and more complex specifications by using schemas as 'building blocks'. This is a feature of the modularity of Z Specifications.

The schema for *CheckRes* can be re-written using the Ξ convention to make it slightly shorter:

―― *CheckRes* ――――――――――――――――――――
ΞReservoir
current_level! : N
RSM! : {Too_Low, Satisfactory, Too_High}
――
(level > max − 250 ⇒ RSM! = Too_High) ∧
(level < min + 100 ⇒ RSM! = Too_Low) ∧
(level ≤ max − 250 ∧ level ≥ min + 100 ⇒ RSM! = Satisfactory) ∧
current_level! = level
――――――――――――――――――――――――――――

In general, it is a very good idea to use the known Z conventions because, for example, when the reader sees the Ξ symbol it is obvious that the operation will not change state and it is not left for the reader to have to scan the schema and work that fact out 'manually'.

The next three sections deal with some fairly important aspects of the Z Notation which may appear complicated to start with. It is advisable to read over it and come back to it later, after seeing more Z, if the material proves hard to follow at first.

## Combining Schemas

We have already seen how useful schema reference can be. Not only does it save us having to write out larger schemas, it also provides an important indication to the reader of the relationships between different schemas in the specification.

One of the strengths of Z is its modularity. We can take a collection of mathematical expressions, put them in a schema, and give them a name. This schema can in turn be combined with other schemas to form another named schema. Thus, if we continue this process, we can build up large, complex structures which help us more naturally model the complex structures encountered in the real world.

One way of combining schemas is by using schema reference. However, there are several other methods of combining them in different ways to suit differing requirements. We can use special **schema operators** to build up **schema expressions**.

At this stage of the book we will only cover the most commonly used forms of schema expressions, later chapters will introduce other forms.

Firstly, we will introduce some Z Notation which can be used to define schemas in a *horizontal* form rather than the usual box-like *vertical* form:

schema name $\hat{=}$ [signature part | predicate part]

where the predicate part is optional as in the vertical form (it defaults to *true* when omitted).

* $S \hat{=} [x : A \mid P]$
  ...denotes the schema $S$ containing an object drawn from set $A$ satisfying $P$.

For example, the state space for the HEP control system could have been defined as follows:

$Reservoir \hat{=} [level, max, min : \mathbb{N} \mid level \geq min \wedge level \leq max]$

Notice the *defines* symbol $\hat{=}$ which is an equals symbol with a caret on top. This symbol is used in the horizontal form of schema expressions as well.

The logical operators $\wedge, \vee, \neg, \Rightarrow$, and $\Leftrightarrow$ can be used on schemas (as opposed to predicates which they are usually used on). The meanings of the schema expressions which they each form are now discussed ($S$ and $R$ are names of schemas, and ⊙ represents one of the four logical operators other than negation):

* $S ⊙ R$
  ...denotes a schema whose signature contains all the declarations in the signature of $S$ and all the declarations in the signature of $R$, and whose predicate part contains the predicate part of $S$ in parentheses followed by ⊙ and then the predicate part of $R$ in parentheses.

*    $\neg S$

...denotes a schema whose signature contains all the declarations in the signature of $S$, and whose predicate part contains the negation of the predicate part of $S$ in parentheses.

## Example 4.12

Suppose we have the following schema definitions:

$S \quad \widehat{=} [x : A \mid x \in B]$
$R \quad \widehat{=} [y : B \mid y \neq \emptyset]$

and the following schemas built by combining schemas:

$S2 \quad \widehat{=} S \wedge R$
$S3 \quad \widehat{=} \neg R$
$S4 \quad \widehat{=} S \Rightarrow S2$

The following are the fully expanded forms of the schemas $S2$, $S3$, and $S4$:

―――――― $S2$ ――――――――――――――――――
$x : A$
$y : B$
――――――――――――――――――――――――――
$(x \in B) \wedge (y \neq \emptyset)$

We can dispense with the brackets if the precedence is obvious.

―――――― $S3$ ――――――――――――――――――
$y : B$
――――――――――――――――――――――――――
$\neg(y \neq \emptyset)$

Note that $\neg(y \neq \emptyset)$ is equivalent to $y = \emptyset$.

―――――― $S4$ ――――――――――――――――――
$x : A$
$y : B$
――――――――――――――――――――――――――
$(x \in B) \Rightarrow (x \in B \wedge y \neq \emptyset)$

Combining schemas can be very useful in a Z specification document. The modularity which comes with it can help in reading and writing large specifications.

For example, the ability to write several schemas and later join them allows us to specify successful operations that can be carried out in a system, then specify what should be done if an error occurs. After having focused on each case we can then join the corresponding schemas together to give a general schema which describes what to do in the event of either a successful or an unsuccessful use of the operations.

## Example 4.13

Consider the following definitions:

```
┌─────── ExampleOp ────────────────────────────
│ ΔExample
│ y? : A
├──────────────────────────────────────────────
│ y? ∉ x ∧
│ x' = x ∪ y?
└──────────────────────────────────────────────
```

```
┌─────── Example ──────────────────────────────
│ x : A
└──────────────────────────────────────────────
```

This operation adds $y?$ to the set $x$ as long as it was not already in the set. If it was in the set then $y? \notin x$ would evaluate to false and so $x' = x \cup y?$ would not be asserted (remember, in conjunction, all predicates need to be true for the whole predicate to be true).

When $y? \in x$, the specification breaks down—it does not say what state the system should be in if an attempt is made to execute *ExampleOp*. We need a specification of what to do when $y? \in x$.

We will define a schema *ErroneousY* and then combine it with *ExampleOp*.

$$\begin{array}{|l}\hline \textit{ErroneousY} \\ \hline \Xi\textit{Example} \\ y? : A \\ \textit{error\_mess!} : MESSAGE \\ \hline y? \in x \land \textit{error\_mess!} = \text{``Input y Already in x''} \\ \hline \end{array}$$

This schema returns an error message when $y? \in x$ is true, and asserts that the state of the system should remain unchanged. Note that the error message in reality should be far more descriptive and meaningful!

What we must do now is join *ExampleOp* and *ErroneousY* to form a new schema which describes what must be done for all possible inputs. We use schema disjunction because only one of the two need evaluate to true:

$NewExampleOp \triangleq ExampleOp \lor ErroneousY$

However, this combined schema does not say what should be assigned to *error_mess!* if the input is correct. We must do this, so we will define the following:

$CorrectY \triangleq [\textit{error\_mess!} : MESSAGE \mid \textit{error\_mess!} = \text{``OK''}]$

So the final, correct, specification of the example operation is as follows:

$FinalExampleOp \triangleq (ExampleOp \land CorrectY) \lor ErroneousY$

Notice that we need schema conjunction to combine *ExampleOp* and *CorrectY* because we specify that both must be true in the same states.

The following is an expanded version of *FinalExampleOp*:

$$\begin{array}{|l}\hline \textit{FinalExampleOp} \\ \hline \Delta\textit{Example} \\ y? : A \\ \textit{error\_mess!} : MESSAGE \\ \hline (y? \notin x \land x' = x \cup y?) \land (\textit{error\_mess!} = \text{``OK''}) \\ \lor \\ (y? \in x \land \textit{error\_mess!} = \text{``Input y Already in x''}) \\ \hline \end{array}$$

# Scope

The variables introduced in Z Specifications have certain restrictions on where they can be used or 'seen'. Each Z variable is associated with a **scope**. The scope of a variable is the area of the specification in which the variable can be used, i.e. where its value can be seen or changed. The variable is said to be **local** to this area.

- The scope of variables declared in the signature of a schema is the predicate part of that schema. However, their scope can be extended to the predicate part of another schema which is combined with them in some way.

- The scope of bound variables in quantification extends as far as the predicate to be quantified, i.e. within the brackets.

There are variables that are visible throughout the rest of the specification after their declaration. These variables are **global** variables.

- The scope of global variables extend from the point of their declaration to the end of the specification document.

### Example 4.14

The following predicate is shown with the scope of the universally quantified bound variable doubly underlined:

$$x > 6 \land (\forall y : A \mid y \in B \bullet y > 6) \land y = 0$$

i.e. the references to the variable $y$ within the brackets are related to the quantification whereas the use of the variable $y$ in $y = 0$ refers to a variable also called $y$ declared elsewhere which is in scope. There could be confusion here so it is best to use different names in such circumstances.

A very important point arising from the above example is that use of a variable refers to the most recently declared value of the variable. For example, suppose we had a global variable $x = 50$. Then, suppose we have a schema which has a variable in its signature also called *x*. The predicate part of that schema (and schemas which refer to the schema) uses the locally declared value of *x*. All points outside the schema (and schemas which refer to it) use the global value of *x*, i.e. 50.

## Axiomatic Descriptions

In a Z specification, there is the idea of a **global signature** and a **global predicate part**. This is a bit like an all-encompassing schema which the whole system model belongs to—it is an extension of the state space. Global variables are declared in the global signature of the specification, and constraints on the specification can be thought of as part of the global predicate part.

All schemas within a specification must satisfy the global predicate part, and the value of global variables declared in the global signature can be used anywhere in the specification after their global declaration without having to be declared again within the schemas referring to them.

### Example 4.15   HEP CONTROL SYSTEM continued...

The state space of the HEP control system introduced a declaration of the variables *max* and *min*. It would be better to declare them in the global signature because their value cannot be changed by any operation on the system and thus they needn't be in the state space schema. We can rewrite the previous schemas to give

─── *Reservoir* ───────────────────
$level : \mathbb{N}$
───
$level \geq min \wedge$
$level \leq max$
──────────────────────────────

─── $\Xi Reservoir$ ───────────────
$\Delta Reservoir$
───
$level' = level$
──────────────────────────────

The above schemas refer to the value of *max* and *min* but they do not attempt to change their values. This is clearer to the reader because previously the reader could have thought that *max* and *min* could be changed by some operations. It also makes dynamic schemas shorter because we don't have to keep on stating

$$max' = max \land min' = min$$

after each operation.

The following Z Notation shows how a global variable may be declared. It is declared in a structure called an **axiomatic description**. This looks rather like a normal Z schema (which declares variables for its predicate part), but the top and bottom lines are removed to give the idea of their global property:

| global signature part |
| --- |
| global predicate part (optional) |

\* The following denotes the declaration of a global variable $x$ of type $A$ which is constrained by the predicate $P$:

| $x : A$ |
| --- |
| $P$ |

The constraint to a global variable can be omitted (it defaults to *true* i.e. when there is no constraint on its values):

\* The following denotes the declaration of a global variable $x$ of type $A$ for which there is no constraint:

| $x : A$ |

i.e. there is simply a vertical bar to the left of the declaration.

\* The following denotes a global constraint $P$ put on a system model:

$P$

i.e. when a predicate appears as a paragraph in a Z specification document then it is a constraint on the values of the global variables previously declared.

### Example 4.16  HEP CONTROL SYSTEM continued...

Near the beginning of the specification of the HEP control system (i.e. before the definition of its state space) there would be the following declaration of the global variables:

$$
\begin{array}{|l}
max, min : \mathbb{N} \\
\hline
max = 3000 \land \\
min = 70
\end{array}
$$

The variables are accompanied by a constraint on their values, i.e. that *max* can only equal 3000 and *min* can only equal 70. This constraint is said to be very **strong**. This is because it restricts the possible values *max* and *min* can take to only one. So, out of all the natural numbers *max* can only equal 3000.

A **weaker** constraint might be *max* $\geq$ 3000 which means that *max* can take on *any* value from the set $\mathbb{N}$ greater than or equal to 3000. Thus a strong predicate makes more restrictions on the number of possible states the predicate can be satisfied by. Weak predicates are satisfied by a larger number of states.

In the above example we may have chosen not to specify the actual values of *max* and *min* at all. This is quite likely in a Z specification since we are not really concerned with such details but are more interested in a higher level abstract view of what the system should do.

In fact, Z Specifications are usually drawn up in an early stage of the production of software (e.g. in the analysis stage or preliminary system design stage) and such details may simply not be known at that point. This high level approach is easier to understand because the reader isn't bogged down with details which are effectively irrelevant to the functionality of the system.

# Basic Types

Another example of approaching the modelling of a system at a high level is by using **basic types**. These are types which are specified as being the elementary types of a model. The contents of the types are not specified. We just declare objects to be of those types without worrying exactly what values they take on.

∗   [A, B, C]

...denotes the introduction of three basic types of a specification called A, B, and C.

### Example 4.17    TELEPHONE DIRECTORY continued...

In the state space of the specification of the telephone directory we declared *teldir* to be of type **P**(*NAME* ×*NUM*). Thus, we implied that there existed a set called *NAME* and a set called *NUM*, but we didn't actually state the contents of those sets. Effectively, *NAME* and *NUM* are basic types of the specification, and the following Z paragraph would appear before any reference to *NAME* and *NUM*:

[*NAME*, *NUM*]

This is a sensible thing to do in this specification because it doesn't really matter what format the name or telephone number are stored as. All one is interested in at the early stages of development is a high-level, abstract view of the data and what operations can be carried out on that data; these are things which do not depend on exactly what a name or number is. Later on we *may* wish to come back and specify in more detail what each type is composed of, if that was thought important. For example:

$$NUM : \mathbf{P}(\mathbf{N} \times \mathbf{N})$$

$$NUM = \{x : \mathbf{N};\ y : \mathbf{N} \mid x < 10000 \wedge y < 10000000\}$$

This states that the type *NUM* is a set of ordered pairs such that the first element is the STD code (which is specified to be no more than four digits), and the second element is the telephone number (which is specified to be no more than seven digits). It is actually asserting that every pair $(x,y)$ in *NUM* must satisfy $(x < 10000 \land y < 10000000)$.

If we changed the specification of *NUM* then it wouldn't affect the specification of the telephone directory. This is because the specification of the telephone directory assumes *NUM* to be an elementary type and thus a building block, whose contents are irrelevant. It is like a house which is built with bricks—we do not care what the bricks are made of precisely, we just know that they are there to be used.

## 4.3 SPECIFICATION LAYOUT CONVENTIONS

A Z specification document consists of 'paragraphs' of Z Notation separated by more informal English paragraphs. The English prose serves to introduce each Z paragraph and explain the meaning of any objects or predicates used which might not be immediately obvious.

Each specification gradually builds up a model of an information system. Types, global variables, and the state space are first declared and are then available for reference by Z paragraphs following on in the document. Later paragraphs typically specify how the state space can be navigated—how the system may legally move from one state to another.

Each Z paragraph should stand out. This is achieved by starting each paragraph on a new line, preferably separated by using a blank line above and below it.

The use of spaces within a Z paragraph is essentially arbitrary. However, the emphasis should always be on a clear presentation of expressions, and not on saving space—although some consideration should be given to the rainforests!

# 4.4 LEARNING BY EXAMPLE

We will now follow an example specification from start to finish using the knowledge gained thus far. The specification is very simple, but it serves to show the layout of a Z specification document, and give a 'feel' for what is to be expected. The informal text surrounding the Z will have a greater explanatory tone than strictly necessary. It is important that an effort is made to understand the Z and to visualise the model in terms of the mathematical entities covered up until now.

## A Simple Computerised Library System

Every time a book comes in to the library, or is lent out, its barcode is read into the computer. At any moment in time the computer is able to tell whether or not a certain book is on loan, reserved, or on the shelves. A reserved book is never on the shelves, it is either on loan or kept under the counter. The library does not keep multiple copies of any book, and only one person can reserve any one book. The library computer can perform the following functions:

i. Add a new book to its database.
ii. Record the details of a book being taken out.
iii. Record the details of returned books collected from a 'returns' box at the end of each day.
iv. Reserve a book.
v. Perform a book query—search for a book in the database to check for its existence. If it does exist in the library, the computer must say whether it is on loan, on the shelves, or reserved.

*A Z Specification of the Computerised Library System*

We can model this system using basic sets. We will have a set called *books* which contains the barcodes of all the books the library owns. Three other sets called *loaned, reserved,* and *shelved* hold the barcodes of the books currently on loan, reserved, or on the shelves of the library respectively. The contents of all these sets represent all the information we are interested in when modelling the library.

We are not interested in the format or make-up of a barcode, and so the type *BARCODE* is a basic type of our model:

[*BARCODE*]

There are some important properties here which can be made explicit to aid understanding. Firstly, the union of the sets *loaned, reserved,* and *shelved* forms the set *books,* i.e. all the books in the library are either on loan and not reserved, on loan and reserved, reserved, or on the shelves. Secondly, as a consequence of the previous statement, a book cannot be on the shelves and reserved at the same time, nor can it be on loan and on the shelves at the same time, i.e. the intersection of *shelved* and *reserved,* and of *shelved* and *loaned* must always be the empty set.

These sets and their properties form a state space with which we can model the computerised library:

```
┌─── Library ────────────────────────────────
│ books : PBARCODE
│ shelved, reserved, loaned : PBARCODE
│ ─────────────────────────────────────────
│ books = loaned ∪ reserved ∪ shelved ∧
│ shelved ∩ reserved = ∅ ∧
│ shelved ∩ loaned = ∅
└────────────────────────────────────────────
```

Initially, the library is empty; it contains no books.

```
┌─── InitLibrary ────────────────────────────
│ Library
│ ─────────────────────────────────────────
│ shelved = ∅ ∧ loaned = ∅ ∧ reserved = ∅
└────────────────────────────────────────────
```

When a new book is delivered, its barcode has to be entered into the computer system and the book must be put on the shelves. This can be modelled by adding that book's barcode to the set *shelved*, but only if the new book is not already in the library. Remember, this simple library does not allow multiple copies of any book.

---
**NewBook**

$\Delta Library$
$bcode? : BARCODE$

---
$bcode? \notin books \wedge$
$shelved' = shelved \cup \{bcode?\} \wedge$
$loaned' = loaned \wedge$
$reserved' = reserved$

---

When a library member wishes to take out a book, the book's barcode is entered into the computer. The barcode must be removed from the set *shelved* or the set *reserved* (if the book was reserved for the member and kept under the counter), and must be added to the set *loaned*:

---
**TakeOutBook**

$\Delta Library$
$bcode? : BARCODE$

---
$bcode? \in (shelved \cup (reserved \setminus loaned)) \wedge$
$shelved' = shelved \setminus \{bcode?\} \wedge$
$reserved' = reserved \setminus \{bcode?\} \wedge$
$loaned' = loaned \cup \{bcode?\}$

---

The predicate $bcode? \in (shelved \cup (reserved \setminus loaned))$ ensures that the barcode entered makes some sense, i.e. that it is the barcode of a book from the shelf or under the counter (a book which is reserved but not on loan)—it wouldn't make any sense if someone tried to take out a book which the library didn't have, or was already out on loan!

Note that the barcode is removed from both *shelved* and *reserved* even though it can only be in one of them. This is okay to do because, for example, the set {1, 2, 3} \ {4} equals {1, 2, 3}, i.e. set difference results in all those members of the first set which happen not to be members of the second set. In other words, set difference doesn't really 'physically' remove certain members, it just excludes the chance for certain members to appear in a set (which might have the effect of removing some members).

While it is useful to think of some operations as adding or taking away objects in a physical sense, it should be borne in mind that we have mathematical tools which are highly abstract and well-defined—we may choose to interpret them one way, but we must never violate their formal definition.

At the end of each day, books from the returns box must be entered into the computer so the computer knows that they are no longer on loan. For each book, the computer tells the librarian whether it should be put on the shelves or under the counter (if it was reserved). This can be modelled by creating a set which holds the barcodes of all those books which must be put on the shelves (added to *shelved*). Of course, the entire contents of the box must be removed from *loaned*.

$$
\begin{array}{|l}
\hline
\quad\quad\textit{Returns} \\
\hline
\Delta Library \\
bcodes?\,:\mathbf{P}BARCODE \\
\hline
bcodes? \subseteq loaned \,\wedge \\
shelved' = shelved \cup \{x : BARCODE \mid x \in bcodes? \wedge x \notin reserved\} \,\wedge \\
loaned' = loaned \setminus bcodes? \,\wedge \\
reserved' = reserved \\
\hline
\end{array}
$$

The set $\{x : BARCODE \mid x \in bcodes? \wedge x \notin reserved\}$ contains precisely the barcodes of those books returned which are not reserved. For the operation to be legal, all the books returned must have been on loan in the first place, hence the precondition $bcodes? \subseteq loaned$.

Reserving a book entails the barcode of the book being entered into the computer (in real life this would be found by a search for the books name or ISBN number for example). If the book is on the shelf or currently already reserved then the librarian is informed, otherwise the book barcode is added to the 'reserves' list (added to *reserved*).

```
┌─ ReserveBook ─────────────────────────────────────────────
│ ΔLibrary
│ bcode? : BARCODE
│ report! : {ok, on_shelves, cur_reserved}
│ ─────────────────────────────────────────────────────────
│ bcode? ∈ books ∧
│ (bcode? ∈ reserved ⇒ report! = cur_reserved ∧ reserved' = reserved) ∧
│ (bcode? ∈ shelved ⇒ report! = on_shelves ∧ reserved' = reserved) ∧
│ (bcode? ∈ loaned \ reserved ⇒ report! = ok
│                  ∧ reserved' = reserved ∪ {bcode?}) ∧
│ loaned' = loaned ∧ shelved' = shelved
└──────────────────────────────────────────────────────────
```

The only time that a book can legitimately be reserved is when it is on loan and not currently reserved, i.e. when its barcode is a member of the set *loaned \ reserved*.

A book query tells the librarian whether a book is in the library, and if it is, then whether it is on loan, on the shelves, or reserved. If it is both on loan and reserved then it is treated as if it were just reserved. We can model this with membership tests.

```
┌─ BookQuery ───────────────────────────────────────────────
│ ΞLibrary
│ bcode? : BARCODE
│ report! : {unknown, shelved, reserved, loaned}
│ ─────────────────────────────────────────────────────────
│ (bcode? ∉ books ⇒ report! = unknown) ∧
│ (bcode? ∈ shelved ⇒ report! = shelved) ∧
│ (bcode? ∈ loaned \ reserved ⇒ report! = loaned) ∧
│ (bcode? ∈ reserved ⇒ report! = reserved)
└──────────────────────────────────────────────────────────
```

The specification given does not yet deal with the illegal inputs, e.g. when an operation is performed on a barcode which doesn't exist. We will say that every operation must output a message after it has taken place. The message must inform the user of any errors:

If an operation is successfully completed then the message will be "operation was successful.".

If the barcode entered was not a barcode of a book in the library, then the message will be "barcode does not match any library book!".

If an attempt is made to add a new book to the library when the barcode of the book is already in the library, then the message will be "no multiple copies of books allowed!".

When the barcodes of some of the returned books are not in the library, then the message will be "some barcodes do not match any library books!". When the barcodes of some of the returned books are not supposed to be on loan, then the message will be "some barcodes are of books not on loan!".

If the barcode entered when attempting to take out a book is already on loan then the message "barcode corresponds to book already on loan!"

We will introduce a new type *LIBMESS* as a global variable. It holds the possible error messages.

$$
\begin{array}{|l}
LIBMESS : \{ \text{"operation was successful."}, \\
\quad \text{"no multiple copies of books allowed!"}, \\
\quad \text{"barcode does not match any library book!"}, \\
\quad \text{"some barcodes don't match any library books!"}, \\
\quad \text{"some barcodes are of books not on loan!"}, \\
\quad \text{"barcode corresponds to book already on loan!"} \}
\end{array}
$$

The following schema corresponds to the successful completion of an operation:

$$
\begin{array}{|l}
\underline{\quad Success \quad} \\
mess! : LIBMESS \\
\hline
mess! = \text{"operation was successful."}
\end{array}
$$

The next four schemas specify the response to an error in one of the four operations other than *BookQuery*:

┌─ *NoMultiples* ─────────────────────────────
│ ΞLibrary
│ bcode? : BARCODE
│ mess! : LIBMESS
│ ─────────────────────────
│ bcode? ∈ books ∧
│ mess! = "no multiple copies of books allowed!"
└─────────────────────────────────────────────

┌─ *TakeOutErrors* ───────────────────────────
│ ΞLibrary
│ bcode? : BARCODE
│ mess! : LIBMESS
│ ─────────────────────────
│ (bcode? ∉ books ⇒ mess! =
│     "barcode does not match any library book!") ∧
│ (bcode? ∈ loaned ⇒ mess! =
│     "barcode corresponds to book already on loan!")
└─────────────────────────────────────────────

┌─ *ReturnErrors* ────────────────────────────
│ ΞLibrary
│ bcodes? : ℙBARCODE
│ mess! : LIBMESS
│ ─────────────────────────
│ (bcodes? ⊄ books ⇒ mess! =
│     "some barcodes don't match any library books!") ∧
│ (bcode? ⊄ loaned ∧ bcodes? ⊆ books ⇒ mess! =
│     "some barcodes are of books not on loan!")
└─────────────────────────────────────────────

┌─ *NotLibraryBook* ──────────────────────────
│ ΞLibrary
│ bcode? : BARCODE
│ mess! : LIBMESS
│ ─────────────────────────
│ bcode? ∉ books ∧
│ mess! = "barcode does not match any library book!"
└─────────────────────────────────────────────

We now give the full specification of each function:

| | |
|---|---|
| *NewBook2* | $\triangleq (NewBook \wedge Success) \vee NoMultiples$ |
| *TakeOutBook2* | $\triangleq (TakeOutBook \wedge Success) \vee TakeOutErrors$ |
| *Returns2* | $\triangleq (Returns \wedge Success) \vee ReturnsErrors$ |
| *ReserveBook2* | $\triangleq (ReserveBook \wedge Success) \vee NotLibraryBook$ |
| *BookQuery2* | $\triangleq BookQuery \wedge Success$ |

## Comments on the Specification

The above Z specification gives a feel of what is to come, but is not very elegant. The mathematics used seemed awkward to use, and more importantly, the notation used up to now limited what we could model. The following points are of what should have been in the above specification if we wanted to model a more realistic library situation:

- Not only should the barcode for each book be kept, but also at least the book's title, its author, and its ISBN number. In fact, storing the *British Library Cataloguing in Publication* data for each book would be a good idea.

- Real libraries have multiple copies of books—there should be no restrictions on this!

- In practice, many people may wish to reserve the same book. Therefore, the computer should implement some sort of queueing system for reservations for each book.

- Each library member should be stored on the computer so that it knows who has got which books on loan, and when they are due back.

- and so on...

The next few chapters will introduce the mathematical notation required to allow the above, and more, to be specified in Z. It enables this to be done in an elegant way so that the awkwardness and limitations will no longer be a problem.

## 4.5 SUMMARY

This chapter has introduced many new concepts. It is of fundamental importance to understand the notion of state. The fact that we can define a state space for a system means we are attempting to capture the very essence of what the system is to achieve. If the system strays from the state space then it is not doing what it should and an error has occurred. The Z Specifications help ensure that this does not happen by highlighting the errors when they arise.

Z Schemas represent a very powerful and important mechanism for modelling large systems. They behave like named packages which assert some static or dynamic property of a system. The property asserted can be included in other schemas by using schema reference and other schema expressions. This brings in benefits of abstraction and modularity. The abstraction helps make it easier to understand a specification while the modularity helps introduce small (more manageable) blocks of Z at a time and then re-use the blocks or make later amendments to the blocks without too much effort.

The foundations have been laid for formally specifying an information system. We will now extend our mathematical vocabulary to enable us to successfully build on what we have so far, and be able to understand real Z Specifications of real-world systems.

# CHAPTER FIVE

# Relations

The mathematical tools that we will cover in the next few chapters will be built from the set theory that we have studied up to now. In this chapter we will introduce some notation which allows us to easily show relationships between objects.

The ability to specify relationships in a system is very important. Many of the real-world entities that we will want to model are related in some way to the other entities in their system; some of these relationships may be immediately obvious, while others may be rather more subtle.

The relationships between the objects represent the bonds that tie what may be a large, complex system together to form a cohesive, structured amalgamate which is an essential property of a modern well-designed integrated computer system. Instead of a system appearing as a soup of independent objects swimming around we will have more of a network of objects with interrelationships explicitly identified.

As an example of a relationship, consider modeling the preparation of a school timetable. In the model the set of teachers in a school can be considered to be related to the set of pupils in the school by the relationship called "teaches". That is, teacher *x* "teaches" pupil *y*. We need some mechanism for expressing such links between objects in Z Specifications.

*Definition*

A **relation** in the Z Notation can be modelled as a set of ordered pairs $(x, y)$ which satisfy some rule which states that, for each pair in the set, object *x* is related to object *y*.

## Example 5.1

Suppose we had a relation called *EarnsMoreThan* which was the set of all pairs $(x,y)$ of employees of a certain firm such that *x* earns more than *y*. Now if Bill and Fred were employees of the firm we could find out if Fred earns more than Bill by finding out if (Fred, Bill) ∈ *EarnsMoreThan*.

If the above is true, then we say that Fred is related to Bill by the relation "earns more than".

*Notation*

There are some special forms of notation which come with relations. Two of them are convenient abbreviations for the notation shown above, i.e. that if two objects, *x* and *y*, are related by the relation *R* then $(x, y) \in R$.

*     *x R y*
    ...denotes the predicate which states that *x* is related to *y* by relation *R*.

That is, *x R y* is equivalent to the predicate $(x, y) \in R$. It is a fact which can be either true or false, but when stated as a constraint then it must always be true.

*     $x \mapsto y$
    ...denotes a **maplet**. It is another way of writing the ordered pair $(x, y)$.

The maplet can provide a useful way of displaying ('mapping out') a relation in a set. The direction of the arrow makes it clear in which way the relation applies. It shows that a relation *R* which contains $x \mapsto y$ "maps *x* to *y*".

Remember that $x \mathrel{R} y$ is equivalent to stating $(x \mapsto y) \in R$.

### Example 5.2

Suppose we have a set of people called *people* of type P*NAME* such that

$people = \{\text{Anna}, \text{Dave}, \text{Alfred}, \text{Gavin}, \text{Elizabeth}\}$

Now, we can define a relation between the objects in this set as follows:

$IsParentOf = \{\text{Dave} \mapsto \text{Gavin}, \text{Anna} \mapsto \text{Gavin}, \text{Alfred} \mapsto \text{Dave},$
$\text{Elizabeth} \mapsto \text{Dave}\}$

This relation asserts that Dave is a parent of Gavin, and so is Anna. That is, the two predicates

Dave *IsParentOf* Gavin

Anna *IsParentOf* Gavin

are both true.

A third form of special notation is the relation symbol $\leftrightarrow$. This is used to denote the type of a relation.

* $A \leftrightarrow B$
  ...denotes the type of a relation between the set $A$ and the set $B$, and is a shorthand for $P(A \times B)$.

When declaring a relation, it is better to show its type using the relation symbol (rather than simply $P(A \times B)$ for a relation between $A$ and $B$) because it makes it immediately obvious to the reader what the set contains and how the ordered pairs are to be treated.

### Example 5.3    UNIVERSITY INFORMATION SYSTEM

A University wishes to develop a computerised information system, called UIS, which keeps track of the following:

i. The names of the students in the University.
ii. The names of the lecturers in the University.
iii. The course code of each of the courses taught at the University.
iv. The names of the lecturers who teach each of the courses.
v. The names of the students who attend each of the courses.
vi. The time of day each course is taught at, and in which room.

The system will be used to provide information for examination registration, examination notification, production of individualised timetables, and for the location of members of staff and students in case of emergencies.

We can use relations to model the UIS system. The names of students are taken from the set *STUDENT*, the names of lecturers are taken from the set *LECTURER*, the course codes are taken from the set *CRSCDE*, and the room numbers are taken from the set *ROOM*. These sets will be used as the basic types of our specification:

[*STUDENT, LECTURER, CRSCDE, ROOM*]

Each day starts at 9am and finishes at 5pm. Every lecture lasts for one hour, and there are seven time slots each day for lectures, i.e. 9am-10am, 10am-11am, 11am-12pm, 12pm-1pm, 2pm-3pm, 3pm-4pm, and 4pm-5pm. There are no lectures between 1pm and 2pm for lunch. The University is open for lectures five days a week.

Each time slot for each week (there are 35 in total) has a unique number from 0 (Monday 9am) through to 34 (Friday 4pm). These numbers are drawn from the natural numbers, but for clarity we will define the following global set:

$PERIOD : \mathbb{N}$

$PERIOD = 0..34$

Now, there are essentially three different relations we can use in order to model the system and its requirements:

- *Attends* maps each student to each course the student attends at University, and is of type $STUDENT \leftrightarrow CRSCDE$.

Thus, if a student attends eight courses, there will be eight maplets in the relation with that student's name on the left and one of the course codes on the right.

- *IsTaughtBy* maps each course to the lecturer teaching that course, hence its type is $CRSCDE \leftrightarrow LECTURER$.
- *IsAt* relates a course to a time period when the course is taught and in which room, i.e. it is of type $CRSCDE \leftrightarrow (PERIOD \times ROOM)$.

So, if a course was lectured three times a week, there would be three maplets in the relation which mapped the course code to each different period and room number.

For example, if the course B43 was held at 10am on Monday in room 212, at 10am on Wednesday in room 212, and at 3pm on Friday in room 241, then the relation *IsAt* would contain the following subset:

$$\{B43 \mapsto (1, 212), B43 \mapsto (15, 212), B43 \mapsto (33, 241)\}$$

# Relational Operators

We have already used relations implicitly. In Chapter Three it was mentioned that = and $\in$ were the basic, elementary predicates of Z, and that all other predicates were derived from them. However, we then proceeded to use the operators $<, >, \leq,$ and $\geq$ as if they were elementary predicates. In fact, they are not. They are relations, and that is one of the reasons why they are often called *relational operators*.

The predicate $x < y$ is an abbreviation for the Z predicate

$$(x, y) \in <$$

where "<" is the 'name' of the relation which relates every integer $x$ to those integers $y$ which are greater than $x$. For example, the set

$$\{..., 1 \mapsto 2, 1 \mapsto 3, 1 \mapsto 4, ..., 2 \mapsto 3, 2 \mapsto 4, ..., 100 \mapsto 101, 100 \mapsto 102, ...\}$$

gives a general idea of the contents of the relation <. Thus, $1 < 2$ is true because $1 \mapsto 2$ is in <, and $1 < 1$ is false because $1 \mapsto 1$ is not in <.

## 5.1 FUNDAMENTAL CONCEPTS

This section deals with the basic concepts and mathematical ideas that go hand-in-hand with relations.

### Domain

If we have a relation $R : A \leftrightarrow B$, then we say that the **domain** of $R$ is the set of all objects $x$ in $A$ such that $x \mapsto y$ for some $y$ in $B$. That is, the domain of a relation is the set of all those elements which appear on the left side of each and every maplet in the relation, and is of type $\mathbf{P}A$. For example, if

$A = \{a, b, c, d, e\}$

$B = \{1, 2, 3\}$

and we had the relation

$R = \{a \mapsto 1, a \mapsto 3, c \mapsto 2, d \mapsto 2\}$

then the domain of $R$ would be $\{a, c, d\}$. It is a subset of $A$.

The notion of relational domain can be expressed with the *dom* operator.

*     dom $R$
  ...denotes the set which is the domain of the relation $R$.

### Example 5.4    UIS SYSTEM continued...

We can produce a set of all students taking courses at the University by extracting the domain of the relation *Attends*. This can be expressed with the axiomatic description

$\quad$ students : STUDENT
$\quad$ ―――――――――――
$\quad$ students = dom Attends

Note that as the *dom* operator results in a set, duplicate student names will be 'lost'. *STUDENT* is a basic type so we can specify that no two objects of this type will be the same. If we later decided to refine the specification we may choose to code the name as an ordered pair of surname followed by other names and hope that no two students have exactly the same pair of names. Another way of getting around duplicates would be to give each student a unique identification number.

The set described using the *dom* operator on a relation $R$ from $A$ to $B$ can be defined using set comprehension as

$$dom\ R = \{x : A \mid (\exists\, y : B \bullet x\ R\ y)\}$$

which means the domain of $R$ is the set of all objects $x$ such that there exists at least one $y$ where $x\ R\ y$ is true. The domain of this relation is of type $\mathbb{P}A$.

## Range

The **range** of a relation $R$ from $A$ to $B$ is the set of all objects $y$ for each maplet $x \mapsto y$ in $R$. The range is of type $\mathbb{P}B$.

∗    $ran\ R$
    ...denotes the set which is the range of relation $R$.

More formally, for a relation $R : A \leftrightarrow B$,

$$ran\ R = \{y : B \mid (\exists\, x : A \bullet x\ R\ y)\}$$

### Example 5.5    UIS SYSTEM continued...

Just as we produced the set of all students by using the domain operator, we can produce the set of all lecturers teaching courses at the University by extracting the range of the relation *IsTaughtBy*. Thus we have

| *teaching* : *LECTURER* |
|---|
| *teaching* = *ran IsTaughtBy* |

Again, we are assuming that each lecturer object is distinguishable.

# Identity

There is a special relation called the **identity relation**. It is the relation which maps every member of a given set to itself. Thus the identity relation on $A$ is the relation which, for each $x$ in $A$, contains the maplet $x \mapsto x$.

*     $id\,A$
  ...denotes the identity relation on $A$.

For example,

$$id\,\{1, 2, 3, 4\} = \{1 \mapsto 1, 2 \mapsto 2, 3 \mapsto 3, 4 \mapsto 4\}$$

## Example 5.6    CAR INSURANCE

A firm of consultants owns a fleet of company cars for its managers. Car insurance is taken out in the name of the principle driver of the car. However, there may be situations when a manager will need to be able to drive another car in the fleet, and would need to be insured to drive that car.

If we had a relation *CanDriveTheCarOf* which mapped a manager to the name of the manager whose car he/she can drive then we can see that at the very least this relation should contain the identity on managers

$$id\,managers \subseteq CanDriveTheCarOf$$

because all managers can drive their own cars!

# Relational Inversion

Sometimes we may wish to reverse the direction a relation applies, i.e. form a new relation by taking each maplet $(x,y)$ in the old relation and putting a corresponding $(y,x)$ in the new relation. This is known as **relational inversion**. For example, the inverse of the relation

$$\{1 \mapsto 2, 2 \mapsto 3, 1 \mapsto 3, 3 \mapsto 4\}$$

is the relation

$$\{2 \mapsto 1, 3 \mapsto 2, 3 \mapsto 1, 4 \mapsto 3\}$$

\* $R\tilde{}$

...denotes the inverse of the relation $R$.

Thus, if we have a relation $R : A \leftrightarrow B$, then $R\tilde{}$ is of type $B \leftrightarrow A$ and can formally be defined as

$$R\tilde{} = \{y : B; x : A \mid x R y \bullet y \mapsto x\}.$$

That is, the inverse of $R$ is the set of maplets from $y$ to $x$ where $xRy$ is satisfied.

### Exercise 5.1

The UIS system contains the relation *Attends* which maps students to the courses they attend. We could have chosen instead to have the relation *IsAttendedBy* which maps the courses to the students.

Define *IsAttendedBy*, and state its type.

## Relational Image

The range of a relation $R : A \leftrightarrow B$ is the set of all objects $y$ of $A$ to which $R$ relates some member $x$ of $B$. However, sometimes we may only be interested in a subset of this range. One sort of subset we may be interested in is the range of a subset of the relation. For example, if we had the relation

$$\{1 \mapsto 5, 2 \mapsto 10, 3 \mapsto 15, 4 \mapsto 20, 5 \mapsto 25, 6 \mapsto 30\}$$

and we were only interested in the subset of the relation whose domain is in the set of even numbers then we get the relation

$$\{2 \mapsto 10, 4 \mapsto 20, 6 \mapsto 30\}$$

The range of this relation is the set $\{10, 20, 30\}$.

The **relational image** of a relation $R : A \leftrightarrow B$ with respect to the set $S$ is the set of all objects $y$ in $B$ such that an object $x$ in $S$ maps to $y$ in the relation $R$. The set formed is a subset of the range of $R$.

* $R(\!(S)\!)$

   ...denotes the set which is the relational image of $R$ with respect to the set $S$.

### Example 5.7    UIS SYSTEM continued...

Suppose there were several lectures which had to be cancelled at a future date. The course codes of the cancelled lectures are held in the set *cancelled*. Now, the students must be notified of the cancellations in advance, so the University wants to send the appropriate students the information.

We can model the set of all students to be notified using the relational image of the inverse of *Attends* with respect to *cancelled*, i.e. the students to be notified are in the set *Attends*$^\sim(\!(cancelled)\!)$

Thus suppose we had the following data in the *Attends* relation (shown as a table as it might be stored in a computer):

| STUDENT | CRSCDE |
|---------|--------|
| Fisher  | B43    |
| Fisher  | C40    |
| Sharp   | B43    |
| Sharp   | C22    |
| Lucas   | C22    |
| Lucas   | C40    |
| Parsley | B20    |

If we have *cancelled* = {B43}, then

    *Attends*$^\sim(\!(cancelled)\!)$ = {Fisher, Sharp}

Formally, for a given relation $R : A \leftrightarrow B$ and a set $S$ of type $\mathbf{P}A$, we can define the following:

$$R(\!(S)\!) = \{y : B \mid (\exists x : A \mid x \in S \land x \, R \, y)\}$$

That is, the relational image of $R$ with respect to $S$ is the set of all objects $y$ in $B$ such that there exists an object $x$ in $A$ which is a member of $S$ and maps to $y$ in relation $R$.

## Classification of Relations

Certain types of relation have special names. A relation whose domain and range are of the same type is said to be **homogeneous**, e.g. a relation $R$ of type $A \leftrightarrow A$ is homogeneous. Thus, the identity on a set is homogeneous.

Some homogeneous relations have further recognised properties. Consider the relation $R : A \leftrightarrow A$ where $R$ is homogeneous:

- $R$ is said to be **reflexive** if it *contains* the identity relation, i.e. if it relates each member of $A$ to itself, that is $x R x$ is true for all $x$ in $A$.
- $R$ is said to be **symmetric** if whenever $x R y$ is true then $y R x$ is also true, where $x$ and $y$ are both in $A$.
- $R$ is said to be **transitive** if whenever $x R y$ and $y R z$ are both true, for some objects $x, y,$ and $z$ in $A$, then $x R z$ is also true.

These terms are not used very often in Z Specifications. However, they can be quite useful in classifying the relations used to model systems—if we are told a relation is reflexive and transitive then we immediately recognise some of its properties which helps us understand the model more quickly.

### Example 5.8    CAR INSURANCE continued...

We could have stated at the beginning of Exercise 5.6 that the relation *CanDriveTheCarOf* was reflexive rather than assert that

   id managers $\subseteq$ CanDriveTheCarOf.

An important property of the relation could thus have been expressed in an understandable shorthand, without having to resort to the constraint.

## 5.2 ADVANCED OPERATIONS ON RELATIONS

The relational operators covered in this section take relations as their operands and produce relations as their results. They can be very useful in a specification which relies heavily on relations and wishes to use them in elaborate ways to mimic complex real-world relationships.

### Relational Composition

It is often very useful to combine two relationships. **Relational composition** is the process of creating a new relation which represents a combined relationship. For example, if we had a relation that maps names to addresses, and another that maps addresses to telephone numbers, then their relational composition would map names to telephone numbers.

It is important that when composing relationships the type of either the domain or range of either is identical to the domain or range of the other, i.e. they must have a type in common. The relation $A \leftrightarrow B$ can be composed with $A \leftrightarrow C$ or $D \leftrightarrow A$ or even $E \leftrightarrow B$ but not with $C \leftrightarrow D$ since there is no type in common. As another example, the relation

$R = \{1 \mapsto 1, 2 \mapsto 4, 3 \mapsto 9, 4 \mapsto 16\}$

composed with the relation

$S = \{1 \mapsto 2, 4 \mapsto 4, 16 \mapsto 1\}$

forms the relation

$\{1 \mapsto 2, 2 \mapsto 4, 4 \mapsto 1\}$

since the 1 in $R$ maps to 1 which maps to 2 in $S$, and the 2 in $R$ maps to 4 which maps to 4 in $S$, and the 4 in $R$ maps to 16 which maps to 1 in $S$. The 3 in $R$ maps to 9, but 9 maps to nothing in $S$ and so isn't in the composition.

*     $R \mathbin{\text{\S}} S$
    ...denotes the relation which is the relational composition of two relations $R$ and $S$.

* $R \circ S$

   ...denotes backwards relational composition, i.e. it is equivalent to $S \mathbin{\raise.5ex\hbox{$\scriptstyle\circ$}\mkern-1mu\raise-.5ex\hbox{$\scriptstyle\circ$}} R$.

Obviously backwards relational composition has the same effect as normal relational composition—the only difference being the *order* of the operands. This is important because $\mathbin{\raise.5ex\hbox{$\scriptstyle\circ$}\mkern-1mu\raise-.5ex\hbox{$\scriptstyle\circ$}}$ must only be used if the type of the range of the first operand is the same as the type of the domain of the second. For example, if $R : A \leftrightarrow B$ and $S : B \leftrightarrow C$ then we must compose $R$ and $S$ together either as $R \mathbin{\raise.5ex\hbox{$\scriptstyle\circ$}\mkern-1mu\raise-.5ex\hbox{$\scriptstyle\circ$}} S$ or as $S \circ R$.

## Example 5.9    UIS SYSTEM continued...

One useful function the UIS system could perform is provide each student with a list of the lecturers teaching his/her courses. An easy way to do this would be using relational composition.

For each student $x$ we could define the set of lecturers teaching $x$'s course as

$$teach\_x = \{y : LECTURER \mid x \, (Attends \mathbin{\raise.5ex\hbox{$\scriptstyle\circ$}\mkern-1mu\raise-.5ex\hbox{$\scriptstyle\circ$}} IsTaughtBy) \, y\}$$

The composition *Attends* $\mathbin{\raise.5ex\hbox{$\scriptstyle\circ$}\mkern-1mu\raise-.5ex\hbox{$\scriptstyle\circ$}}$ *IsTaughtBy* effectively maps students to lecturers, so the set comprehension above picks out all those lecturers who teach student $x$. Note that the type of *Attends* is $STUDENT \times CRSCDE$, while the type of *IsTaughtBy* is $CRSCDE \times LECTURER$, so normal relational composition is used to match the two $CRSCDE$ types.

Thus if we take the data from before and we had the following data in the *IsTaughtBy* relation:

| CRSCDE | LECTURER |
|---|---|
| B43 | Prof.Shallcross |
| B20 | Prof.Shallcross |
| C22 | Dr.Fox |
| C40 | Dr.Patel |

Then the composition *Attends* $\mathbin{\raise.5ex\hbox{$\scriptstyle\circ$}\mkern-1mu\raise-.5ex\hbox{$\scriptstyle\circ$}}$ *IsTaughtBy* is:

| STUDENT | LECTURER |
|---|---|
| Fisher | Prof.Shallcross |
| Sharp | Prof.Shallcross |
| Parsley | Prof.Shallcross |
| Lucas | Dr.Fox |
| Sharp | Dr.Fox |
| Fisher | Dr.Patel |
| Lucas | Dr.Patel |

If we have $x$ = Fisher, then $teach\_x$ = {Prof.Shallcross, Dr.Patel}.

## Exercise 5.2

For each lecturer $x$, define the set of students who attend that lecturer's course.

## Iteration

It is possible to compose a relation with itself, but of course only if the relation is homogeneous. For $R : A \leftrightarrow A$, $R \mathbin{\raise.3ex\hbox{$\scriptscriptstyle\circ$}} R$ is such a relation. For example, the relation

$\{1 \mapsto 1, 2 \mapsto 4, 3 \mapsto 9, 4 \mapsto 16\}$

when composed with itself is

$\{1 \mapsto 1, 2 \mapsto 16\}$

because 1 maps to 1 which maps to 1, and 2 maps to 4 which maps to 16, but 3 and 4 map to 9 and 16 respectively which both map to nothing.

Sometimes we may wish to extend this idea and compose a relation with itself several times. For example, $R$ composed with itself four times is equal to $((R \mathbin{\raise.3ex\hbox{$\scriptscriptstyle\circ$}} R) \mathbin{\raise.3ex\hbox{$\scriptscriptstyle\circ$}} R) \mathbin{\raise.3ex\hbox{$\scriptscriptstyle\circ$}} R$. This is called relational **iteration**.

* $R^k$
    ...denotes the relation $R$ composed with itself $k$ times.

Thus, $R^2 = R \mathbin{;} R$, $R^3 = (R \mathbin{;} R) \mathbin{;} R$, etc. Note that sometimes relational inversion is written as $R^{-1}$.

### Example 5.10

Suppose we have the relation *IsParentOf* from Example 5.2. The relation *IsParentOf*$^2$ is equivalent to a relation *IsGrandParentOf* since it relates each parent to the child of their child. Thus

$$IsParentOf^2 = \{\text{Alfred} \mapsto \text{Gavin}, \text{Elizabeth} \mapsto \text{Gavin}\}$$

Similarly, *IsParentOf*$^3$ is equivalent to the relation *IsGreatGrandParentOf* etc...

## Closures

We can take a homogeneous relation $R$ and create a new larger relation which is a superset of $R$ (i.e. contains all the maplets in $R$, and more).

The **transitive closure** of a relation $R$ is the smallest possible superset of $R$ with added maplets which ensure that the superset relation is transitive. For example, if

$$R = \{2 \mapsto 3, 3 \mapsto 4, 4 \mapsto 2, 5 \mapsto 2\}$$

then we need to add the maplets $2 \mapsto 4$, $3 \mapsto 2$, $4 \mapsto 3$ and $5 \mapsto 3$ to $R$ to make the relation transitive with the smallest number of additions.

The **reflexive-transitive closure** of a relation $R$ is the smallest possible superset of $R$ with added maplets which ensure that the superset relation is both transitive and reflexive. For example, given

$$R = \{a \mapsto j, f \mapsto g, g \mapsto g, j \mapsto a, k \mapsto f\}$$

then we need to add the maplets $a \mapsto a$, $j \mapsto j$, and $k \mapsto g$ to make the relation transitive, and the maplets $a \mapsto a$, $f \mapsto f$, $j \mapsto j$, and $k \mapsto k$ to make the relation reflexive. Thus the reflexive-transitive closure of $R$ is the relation

$\{a \mapsto j, f \mapsto g, g \mapsto g, j \mapsto a, k \mapsto f, a \mapsto a, f \mapsto f, j \mapsto j, k \mapsto k, k \mapsto g\}$.

The following Z Notation is used to represent the closure operators:

* $R^+$
  ...denotes the relation which forms the transitive closure of $R$.

* $R^*$
  ...denotes the relation which forms the reflexive-transitive closure of $R$.

## Exercise 5.3

Another possible definition of transitive closure for a relation $R : A \leftrightarrow A$ is

$$R^+ = R \cup \{x, z : A \mid (\exists y : A \mid x\ R\ y \wedge y\ R\ z\ )\}$$

because the transitive closure of $R$ contains all those maplets in $R$ as well as all those maplets in the set construction (which represent the maplets needed to uphold the transitive property of $R$).

Express the reflexive-transitive closure for $R$ in a similar way.

## Example 5.11

Suppose we had the relation *ReportsTo* which mapped each employee in a company to those employees directly above them in the management hierarchy of the firm. So, for example, if Dan was on a particular section in a shop and Becca was the assistant section manager, then

Dan *ReportsTo* Becca.

If Jo was the section manager, then

Becca *ReportsTo* Jo.

It is helpful in such a model to emphasise the hierarchical nature of this homogeneous relation. The relation must have transitive members, i.e. given the above two maplets, the maplet

Dan *ReportsTo* Jo

must also be true. Also, the relation must not contain any reflexive members, e.g.

Dan *ReportsTo* Dan

is nonsensical. Thus we would state the following:

$ReportsTo' = ReportsTo^+ \setminus id\, ReportsTo$

This asserts that the relation *ReportsTo* contains transitive maplets but not any reflexive maplets.

# 5.3 MANIPULATING RELATIONS

There may be instances when we have a large relation which holds some specified information and we want to view only a subset of the information.

In this section we will introduce four relational operators in the Z Notation. Each are similar in that they take a relation $R$ and a set $S$ and form a new relation which is a subset of $R$. Obviously, the set $S$ decides which maplets are left in/taken out of $R$ to create the resulting relation.

## Domain Restriction

This first method of producing a subset of a relation $R$ is to restrict the domain of the relation to include only those objects which we are interested in, held in a set $S$. Thus, the new subset relation is comprised of all those maplets $x \mapsto y$ of $R : A \leftrightarrow B$ which have $x$ in $S$, i.e. it is equal to the relation

$\{x : A; y : B \mid x R y \wedge x \in S\}$

Note that the set $S$ is of type $\mathbb{P}A$ and needs to contain a subset of *dom A* if the resulting relation is to contain any maplets.

For example, if we had a model of a telephone directory, as in the previous chapter, but where *teldir* was a relation from *NAME* to *NUM*, then we could choose to produce a set of telephone numbers of all those people who, say, were members of the local golf club (their names being held in the set *golfclub*). The corresponding numbers will be held in the range of the subset relation formed by **domain restriction** on *teldir* with respect to *golfclub*.

∗     $S \triangleleft R$

...denotes the relation which is the domain restriction of $R$ with respect to $S$ i.e. the subset of $R$ whose domain is a subset of $S$.

### Example 5.12    UIS SYSTEM continued...

The UIS system can be used to provide individualised timetables for students. For example, if we had a set of eight course codes of the courses a particular student attends called *stdcourses*, then we could produce the relation that mapped those courses, and those courses only, to the period of study and room number they are taught in. This can be achieved by using domain restriction as follows:

$StdTimetable = stdcourses \triangleleft IsAt$

The relation *StdTimetable* above would consist of maplets which map each course to the location and time of one of the lectures in the course. From this information it would be relatively easy to draw up an individualised timetable for that student.

### Example 5.13

The following are examples of domain restriction:

- $\{2, 4\} \triangleleft \{1 \mapsto h, 2 \mapsto e, 3 \mapsto l, 4 \mapsto l, 5 \mapsto o\} = \{2 \mapsto e, 4 \mapsto l\}$
- $\{a, b, c\} \triangleleft \{d \mapsto a, e \mapsto b, f \mapsto c\} = \{\}$
- $\emptyset \triangleleft \{h \mapsto 1, e \mapsto 1, l \mapsto 2, o \mapsto 1\} = \{\}$

## Range Restriction

**Range restriction** is similar to domain restriction, except that here we are restricting our *range* of interest. For example, in a library if we had a relation from book titles to the number of copies of each book and we wanted a list of all those books with two or three copies, then we would restrict the range of the relation to the set {2, 3}.

**✱**   $R \triangleright S$

...denotes the relation which is the range restriction of $R$ with respect to $S$ i.e. the subset of $R$ whose range is a subset of $S$.

For a relation $R : A \leftrightarrow B$ and set $S$ of type $PB$, we define

$R \triangleright S = \{x : A; y : B \mid x R y \wedge y \in S\}$.

### Example 5.14

The following are examples of range restriction:

- $\{1 \mapsto h, 2 \mapsto e, 3 \mapsto l, 4 \mapsto l, 5 \mapsto o\} \triangleright \{h, l\} = \{1 \mapsto h, 3 \mapsto l, 4 \mapsto l\}$
- $\{a \mapsto a, b \mapsto b, c \mapsto b, d \mapsto a\} \triangleright \{a, b\} = \{a \mapsto a, b \mapsto b, c \mapsto b, d \mapsto a\}$

### Example 5.15   UIS SYSTEM continued...

A lecturer $x$ may want a list of all the students who attend his/her lectures. This can be modelled by using range restriction:

*attendees* = $dom$ (*Attends* $\triangleright$ $dom$ (*IsTaughtBy* $\triangleright$ $\{x\}$))

*IsTaughtBy* $\triangleright\{x\}$ represents the relation which maps each course $x$ teaches to $x$, thus its domain is the set of all the courses $x$ teaches. When *Attends* is range restricted to this set we have the relation which maps each student $x$ teaches to the courses $x$ teaches. The domain of this relation is the set of all the students that $x$ teaches.

Note that we had to put $x$ in brackets because $x$ is the name of a lecturer but we need a set for restriction operations. By putting $x$ in brackets we are temporarily creating a singleton set.

Suppose $x$ = Prof.Shallcross, then

*IsTaughtBy* $\triangleright\{x\}$ = {B43,B20}

and from the *Attends* relation we can see that

$$Attends \triangleright \{B43, B20\} = \{\texttt{Fisher} \mapsto B43, \texttt{Sharp} \mapsto B43, \texttt{Parsley} \mapsto B20\}.$$

## Relational Subtraction

The restriction operators of the last two sub-sections resulted in a subset relation which only included those maplets of interest. A slightly different, but useful approach is to form a subset relation which *excludes* maplets which we are not interested in. This can be achieved with **domain subtraction** and **range subtraction** operators.

* $S \triangleleft R$

    ...denotes the relation which is the domain subtraction of $R$ with respect to $S$.

* $R \triangleright S$

    ...denotes the relation which is the range subtraction of $R$ with respect to $S$.

### Example 5.16

The following are examples of domain and range subtraction:

- $\{2, 4\} \triangleleft \{1 \mapsto h, 2 \mapsto e, 3 \mapsto l, 4 \mapsto l, 5 \mapsto o\} = \{1 \mapsto h, 3 \mapsto l, 5 \mapsto o\}$
- $\{1, 2, 3\} \triangleleft \{4 \mapsto 1, 5 \mapsto 2, 6 \mapsto 3\} = \{4 \mapsto 1, 5 \mapsto 2, 6 \mapsto 3\}$
- $\emptyset \triangleleft \{h \mapsto 1, e \mapsto 1, l \mapsto 2, o \mapsto 1\} = \{h \mapsto 1, e \mapsto 1, l \mapsto 2, o \mapsto 1\}$
- $\{1 \mapsto h, 2 \mapsto e, 3 \mapsto l, 4 \mapsto l, 5 \mapsto o\} \triangleright \{h, l\} = \{2 \mapsto e, 5 \mapsto o\}$
- $\{1 \mapsto 1, 2 \mapsto 2, 3 \mapsto 2, 4 \mapsto 1\} \triangleright \{1, 2\} = \{\}$

### Exercise 5.4

If we have a relation $R : A \leftrightarrow B$ and a set $S$ of type $\mathbb{P}A$ then the following is a definition of domain subtraction:

$$S \triangleleft R = \{x : A; y : B \mid x \, R \, y \wedge x \notin S\}$$

Express a definition of range subtraction for $R$ and $S$ in a similar way.

### Exercise 5.5

Define the subtraction operators using the restriction operators. The ability to do this shows that we could choose not to use the subtraction operators in a specification. However, they are a convenient shorthand which tell the reader that we wish not to consider members from a certain set at a glance; the reader does not have to determine this by scrutinising the specification.

## 5.5 SUMMARY

Relations are very useful in Z Specifications. By showing relationships between entities, the reader will be able to quickly build a mental image of a system and how each component in the system interact and relate to one another. This is very important in large specifications. In fact nowadays a task called *Entity-Relationship modelling* is often carried out by Systems Analysts to identify the objects in a system and their interrelationships.

Relations are also used in other areas of computing. Powerful business databases are often built on *Relational Databases*. These are databases which logically contain their data in tables; fetching data is a matter of looking at those rows of the table which have values in their specified columns which satisfy some search criteria. The UIS examples in this chapter were actually small relational databases.

We have built up a collection of new notation, all of which was based on sets and logic. This new notation can now be built upon further to give us more descriptive power using more mathematical notation. The next few chapters will provide this extra notation, most of which is directly connected to relations.

# CHAPTER SIX

# Functions

This chapter deals with a special kind of relation called a **function**. It can be used to model those data structures for which an object is only associated with at most one other object, and never any more. For example, one theatre ticket should be associated with precisely one seat in the theatre. Similarly, the square of any given natural number $x$ will always be the natural number $y$ and never any other number, so squaring is a function. However, the square root of a number $x$ is $+y$ and $-y$, and so cannot directly be modelled as a function.

*Definition*

A function $f$ from $A$ to $B$ is a relation which maps an object in $A$ to at most one object in $B$, i.e. a function cannot relate an object in the domain to more than one object in the range.

Recall that a relation is a set of maplets. A function is a relation where every maplet has a different member of the domain on its left hand side, i.e. if we take every $x \mapsto y$ in a function $f$, then all the $x$'s would be different.

This means that functions cannot model one-to-many type relationships, e.g. if a function was used to model the relation *IsParentOf*, then it would imply that a person could only ever have one child—something which is surely wrong!

## Example 6.1

In a payroll system, it is essential that the same person doesn't get paid more than once! A model of such a system could emphasise this by using a function *Earns* which maps an employee's National Insurance (NI) number to that employee's earnings. Each employee has exactly one NI number, and this number is unique to the employee, thus it is not possible for the employee to get paid two wages.

To further emphasise the difference between a general relation and a function, let us look at a mapping graphically. Figure 6.1 shows the relation

$$\{a \mapsto 1, a \mapsto 2, b \mapsto 2, b \mapsto 3\}$$

using Venn diagrams, with the straight lines showing a mapping.

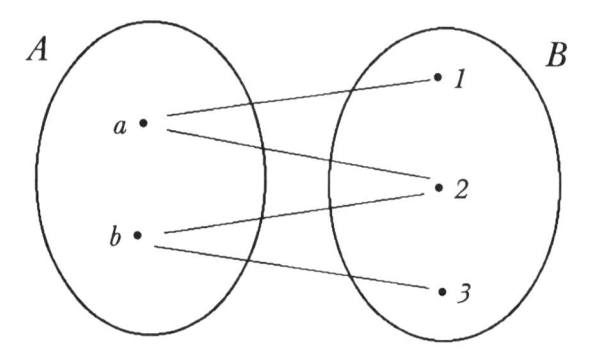

**Figure 6.1**

Note how one object in the domain can relate to more than one object in the range. Such a data structure is very flexible. However, if we know that one object maps to just one other object, we will gain by restricting the data type and enforcing its functional nature. In this way we know more about the nature of the data we wish to model, and can use specialised operators to act on it.

Figure 6.2 shows the function

$$\{a \mapsto 1, b \mapsto 2, c \mapsto 2\}$$

in Venn diagram form. Notice how each object in the domain never associates to more than one object in the range.

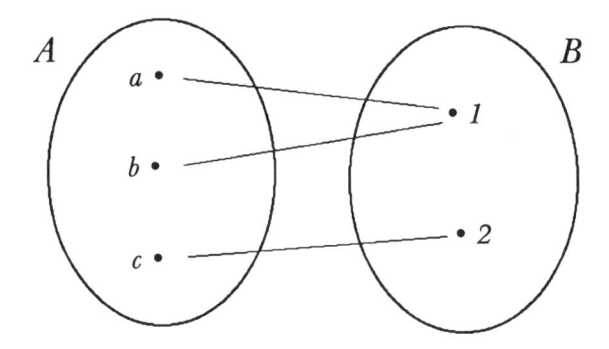

**Figure 6.2**

# 6.1 FUNDAMENTAL CONCEPTS

Since a function is a special kind of a relation, it is possible to use all of the relational notation of Chapter Five and apply it to functions. However, there is some additional notation which can be used with functions to exploit their specialised properties.

## Function Application

If we have a function $f : A \leftrightarrow B$, and we have an $x$ which is a member of $A$ and is in *dom f*, then there exists precisely one maplet $x \mapsto y$ in $f$ for some $y$ in $B$. To put it another way, there is exactly one $y$ in $B$ which is mapped to by $x$. This allows us to use a notational shorthand known as **function application**. This is a process of taking a function and an object in its domain, and using these to find the associated $y$ in the range of $f$ such that $(x \mapsto y) \in f$.

*     $f x$
        ...denotes function application.

The object $x$ in the expression $f\,x$ is often called an **argument** of the function.

Sometimes the argument of a function is enclosed in brackets. Thus, $f\,x$ is equivalent to $f(x)$.

## Example 6.2

Suppose we have the function

$$f = \{1 \mapsto 1,\ 2 \mapsto 4,\ 3 \mapsto 9,\ 4 \mapsto 16,\ 5 \mapsto 25\}.$$

Notice that this function takes an argument and results in a value which is the square of the argument.

Thus, $f\,2$ evaluates to 4, and $f\,5$ evaluates to 25, and we can write such expressions as $f\,5 = 25$ etc.

Function application "associates to the left". This means that if we write

$$f\,x\,y$$

(a function with two arguments) then we mean

$$(f\,x)\,y$$

i.e. $f\,x$ is calculated first (its result must be a set of maplets), and the object $y$ is then applied to the set of maplets to produce the final result of the expression. That is, $f\,x$ should evaluate to a set which is a function which can then be applied to $y$. Note that this is different to a function applied to one argument which happens to be an ordered pair.

## Partial Functions

The functions we have been talking about up to now have been **partial functions**. This is when the domain of the function can be a proper subset of the type of the domain. That is, the domains we have dealt with up until now have not been equal to the set which the domain members were drawn from (the domain type).

For example, if we have a function from the set of all possible book ISBN codes to the set of book information, then in a model of a library system the function would be a partial function. This is because the library could not possibly contain every single book, and hence every single ISBN code (there are many more ISBN codes to be allocated to books that haven't been written yet!). Thus, the set of book ISBN codes which the library does own will be a proper subset of the set of all possible such codes.

*     $A \nrightarrow B$
    ...denotes the set of partial functions from $A$ to $B$.

Thus, the above notation is used to specify the type of a partial function, and is a subset of $\mathbf{P}(A \times B)$.

## Example 6.3    OPTICIANS

An optician keeps records on her patients in a personal computer database. A model of the system uses a function called *record* which maps each patient's computer number (of type *PNUM*) to a set which contains information about the patient (of type *INFO*).

A further function *name* maps each computer number in the system to the name of a patient (of type *NAME*). The state space is as follows:

---
*Opticians*
---
$record : PNUM \nrightarrow INFO$
$name : PNUM \nrightarrow NAME$
---
$dom\ record = dom\ name$
---

The invariant states that every patient in the system has a record. We used partial functions because not all members of *PNUM* are associated with a patient at all times, e.g. when the practice first opens there might be no patients!

## Total Functions

There are some situations where the functions we use in our model map *every* member of the domain type to the range. For example, if we had a function $f : \mathbb{N} \twoheadrightarrow \mathbb{N}$ which mapped every natural number to its square, then the function would be called a **total function**. This is because, for every $x$ in $\mathbb{N}$, there is an $f x$ in $\mathbb{N}$ which is the square of $x$.

* $A \rightarrow B$

   ...denotes the set of total functions from $A$ to $B$, i.e. functions which relate each and every member of $A$ to exactly one member of $B$.

Thus, the above notation is used to specify the type of a total function, and is a subset of $\mathbb{P}(A \times B)$.

Note that

$$A \rightarrow B = \{f : A \twoheadrightarrow B \mid \text{dom} f = A\}.$$

That is, the set of total functions from $A$ to $B$ is equal to the set of partial functions from $A$ to $B$ where the domain is equal to the set $A$.

# 6.2 OPERATIONS ON FUNCTIONS

Many of the operations which can be performed on functions are the same operations used on ordinary relations. However, there is one special operation unique to functions. This will be covered first before going on to discuss the use of the relation operators with functions as arguments.

## Function Overriding

We can use functions to model some store of data which may change or be added to. However, imagine a state of a system where we have a function

$$f = \{a \mapsto 23, b \mapsto 56, c \mapsto 19, d \mapsto 67\}$$

and we wish to change the state by changing the value associated with $c$ to 20. The operation

$$f' = f \cup \{c \mapsto 20\}$$

would not be correct. This is because $f$ is a function and so $c$ can only map to one value. If the above operation was carried out then we would not know whether $f\,c = 19$ or $f\,c = 20$. If we wish $f\,c$ to change to 20 then we would have to override the old value of $c$ with the new one, which can be achieved with the operation

$$f' = (\{c\} \mathbin{\lhd\!\!\!-} f) \cup \{c \mapsto 20\}$$

This ensures that the maplet $c \mapsto 20$ is added to the function $f$ and the maplet $c \mapsto 19$ is deleted from $f$, thus allowing no confusion.

To help clarity in such situations, we can use a special operation known as **function overriding**.

* $f \oplus g$
  ...denotes the function which is the function $f$ overridden with the function $g$.

The function $f \oplus g$ contains all those maplets in $f$ which have domains that do not intersect with the domain of $g$, plus all the maplets in $g$. Note that $f$ and $g$ must be of the same type, e.g. $\mathbb{P}(A \times B)$. A formal definition for function overriding given two functions $f, g$ of type $A \nrightarrow B$ is

$$f \oplus g = ((\operatorname{dom} g) \mathbin{\lhd\!\!\!-} f) \cup g$$

That is, $f \oplus g$ equals the domain subtraction of $f$ with respect to the domain of $g$ (which is the set of maplets in $f$ which do not 'clash' with any maplets in $g$) in union with the function $g$.

## Example 6.4

The following are examples of function overriding:

- $\{1 \mapsto b, 2 \mapsto e, 3 \mapsto l, 5 \mapsto o\} \oplus \{1 \mapsto h, 4 \mapsto l\}$
  $= \{1 \mapsto h, 2 \mapsto e, 3 \mapsto l, 4 \mapsto l, 5 \mapsto o\}$

- $\{a \mapsto 10, b \mapsto 20, c \mapsto 30\} \oplus \{a \mapsto 10, b \mapsto 10, c \mapsto 10\}$
  $= \{a \mapsto 10, b \mapsto 10, c \mapsto 10\}$
- $\{a \mapsto c, d \mapsto f, g \mapsto g\} \oplus \varnothing = \{a \mapsto c, d \mapsto f, g \mapsto g\}$

## Relational Operations

The following operators, introduced in the previous chapter, can be applied to functions and result in a function:

- The restriction operators
- The subtraction operators
- The identity relation

Additionally, the following operators can be applied to functions, but will not necessarily result in a function:

- Relational inversion
- Relational composition

The above two relation operators and some of the set operators such as set union need to be used cautiously when applied to functions since their result may not be a legal function. This may or may not be desirable depending on the system specified. If a legal function is required then we must be certain that the preconditions for the operation eliminate all those cases where the result would be illegal.

### Example 6.5   OPTICIANS continued...

We must model three frequent operations on the Opticians system. The first is the deletion of patients from the system when they move opticians. The second involves the addition of new patients to the system. Finally, the third is the updating of information on the system.

Adding a new patient $x$, with record $r$ and patient number $n$ can be achieved using simple set union operators:

┌─ *AddPatient* ─────────────────────────────
│ $\Delta Opticians$
│ $n? : PNUM$
│ $r? : RECORD$
│ $x? : NAME$
├────────────────────────────────────────────
│ $n? \notin record \land$
│ $record' = record \cup \{n? \mapsto r?\} \land$
│ $name' = name \cup \{n? \mapsto x?\}$
└────────────────────────────────────────────

Updating information kept in the records about a particular patient with patient number $n$ to a new record $r$ will need function overriding. This ensures that the new information overwrites the old information:

┌─ *UpdateRecords* ──────────────────────────
│ $\Delta Opticians$
│ $n? : PNUM$
│ $r? : RECORD$
├────────────────────────────────────────────
│ $record' = record \oplus \{n? \mapsto r\} \land$
│ $name' = name$
└────────────────────────────────────────────

Deleting a patient from the records can be modelled using the domain subtraction operator. Thus, to remove patient number $n$ we have:

┌─ *DelPatient* ─────────────────────────────
│ $\Delta Opticians$
│ $n? : PNUM$
├────────────────────────────────────────────
│ $record' = \{n?\} \mathbin{\lhd\!\!\!-} record \land$
│ $name' = \{n?\} \mathbin{\lhd\!\!\!-} name$
└────────────────────────────────────────────

This means that the function *record'* contains all the maplets in *record* without the maplet for patient number $n$.

# 6.3 SPECIAL TYPES OF FUNCTION

There are some functions which have certain identifiable properties, just like some relations have the properties of transitivity, reflexivity, or symmetry. These functions can also be put into named categories.

In a specification it is not essential to explicitly state if a function conforms to one of the three categories, but again it is best to make such knowledge explicit in order to help the reader understand the model.

## Injections

An **injection** is a special type of function. It is a function in which each member of the domain maps to a unique value in the range, i.e. no member in the range of an injection is mapped to by more than one member of the domain.

For example, if we had a function which mapped names to account numbers for non-joint accounts in a banking system, then the function would be an injection—we certainly wouldn't want two strangers to be associated with the same bank account number!

*     $A \rightarrowtail\!\!\!\rightarrow B$
  ...denotes the set of partial injections from $A$ to $B$.

*     $A \rightarrowtail\!\!\!\rightarrow B$
  ...denotes the set of total injections from $A$ to $B$.

Thus, an injection from $A$ to $B$ is of type $A \rightarrowtail\!\!\!\rightarrow B$ (if it is partial) and this type is a subset of the powerset $\mathbf{P}(A \times B)$.

A more formal definition of the property of a partial injection $f: A \rightarrowtail\!\!\!\rightarrow B$ is

$$(\forall x, y : A \mid x \in dom f \wedge y \in dom f \bullet f x = f y \Rightarrow x = y)$$

This states that when we take any two members from the domain of $f$, if the predicate $f x = f y$ is satisfied then it must be that $x$ is the same object as $y$.

# Surjections

A **surjection** from $A$ to $B$ is another special type of function, but this time where the range is equal to $B$, i.e. every member of $B$ is mapped to by one or more members of the domain of the function.

### Example 6.6

Suppose we were to model a scheduling system which allocated several jobs to employees in a firm. We could use a surjection from the set of jobs to the set of employees. We need a surjection because we must ensure that everyone gets at least one job!

* $A \nrightarrow\!\!\!\rightarrow B$
    ...denotes the set of partial surjections from $A$ to $B$.

* $A \rightarrow\!\!\!\rightarrow B$
    ...denotes the set of total surjections from $A$ to $B$.

Thus, a surjection from $A$ to $B$ is of type $A \nrightarrow\!\!\!\rightarrow B$ (if it is partial) which is a set that is a subset of the powerset $\mathbf{P}(A \times B)$.

A more formal definition of the property of a partial surjection $f: A \nrightarrow\!\!\!\rightarrow B$ is

$$(\forall y : B \mid (\exists x : \text{dom} f \bullet f x = y))$$

This states that, for every member $y$ of $B$, there exists at least one $x$ in the domain of $f$ such that $f x = y$.

# Bijections

The last special function we will consider is a combination of the concepts of injectivity and surjectivity. A **bijection** is a total function from $A$ to $B$ in which each member of $A$ is mapped to a different member of $B$, and each member of $B$ is mapped to by exactly one member of $A$.

* $A \rightarrowtail\!\!\!\rightarrow B$
    ...denotes the set of bijections from $A$ to $B$.

Thus, a bijection from $A$ to $B$ is of type $A \rightarrowtail\!\!\!\rightarrow B$ and this type is a subset of the powerset $\mathbf{P}(A \times B)$.

## Example 6.7

Sometimes we may wish to model a structure in which a collection of objects are related to another collection of objects in one-to-one correspondence. A generic example of this sort of structure is a distribution-type system where $n$ 'balls' must be distributed to $n$ 'containers'.

## Example 6.8

If $A = \{1, 2, 3\}$ and $B = \{a, b, c\}$, then

$$A \rightarrowtail\!\!\!\rightarrow B = \{\{1 \mapsto a, 2 \mapsto b, 3 \mapsto c\}, \{1 \mapsto b, 2 \mapsto c, 3 \mapsto a\},$$
$$\{1 \mapsto c, 2 \mapsto a, 3 \mapsto b\}, \{1 \mapsto c, 2 \mapsto b, 3 \mapsto a\},$$
$$\{1 \mapsto a, 2 \mapsto c, 3 \mapsto b\}, \{1 \mapsto b, 2 \mapsto a, 3 \mapsto c\}\}.$$

Thus, any function from $A$ to $B$ which is a bijection is a member of the above set.

Another definition of the property of a bijection $f : A \rightarrowtail\!\!\!\rightarrow B$ is as follows:

$$(\forall y : B \bullet (\exists_1 x : A \bullet f x = y))$$

This states that, for every member $y$ of $B$, there exists exactly one $x$ in $A$ such that $f x = y$.

## 6.4 SUMMARY

Many information systems have structures which lend themselves to modelling using functions. A function takes one object and transforms it into another object in a *deterministic* way, i.e. when we have a particular object $x$, if $f x = y$ then $x$ will always be transformed to $y$ and no other object.

In a relation, if we had an object $x$ it could be related to several other objects and not any one in particular.

All functions are either partial or total. Additionally, there are some terms for special types of functions which, when used, give the reader more of a clue about the relationship between the objects in the domain and those in the range. Giving readers such clues helps them to gain understanding of the structure and the data in a shorter time.

# CHAPTER SEVEN

# Sequences

The mathematical set provides a useful mechanism for modelling stores of information. However, we are limited to testing for membership and cardinality. The objects are either in a set or out of a set. There is no concept of order or repetition—we cannot say that one object comes before another, or that a certain object occurs $n$ times. This chapter introduces the notion of a **sequence** of objects which captures both order and repetition information for the data modelled.

*Definition*

A sequence is an *ordered* list of objects. A sequence $s$ of objects of type $A$ is modelled by a partial function $s : \mathbb{N} \nrightarrow A$ from the set of natural numbers to the set of objects we wish to order.

What we do is give a number to each object which reflects its order; the first object is mapped to by 1, the second object is mapped to by 2, and so on.

For example, the sequence "2, 3, 3, 6" contains the elements 2, 3, 3, and 6, in that order. The second element is 3 and the last is 6.

*Notation*

Sequences are normally displayed within angled braces, i.e. between " ⟨ " and " ⟩ ". However, a big difference between the set display and sequence display mechanisms is that the order of the objects in a sequence display is very important. We must display the elements of a sequence as an order-preserving list of objects.

**Example 7.1    WORD PROCESSOR**

We could model a word processor document as a sequence of letters, numbers, and punctuation. At the beginning of some correspondence, the first few letters might form the word "Dear". This sequence of letters could be displayed as

⟨D, e, a, r⟩

Note that we could not write the sequence in any other way. That is,

⟨D, e, a, r⟩ ≠ ⟨D, a, e, r⟩ ≠ ⟨D, e, e, a, a, a, r⟩.

# 7.1 FUNDAMENTAL CONCEPTS

### Sequence Types

A special type shorthand exists for sequences which should be used whenever a sequence is declared in a Z Specification.

* *seq A*
    ...denotes the set of all sequences of objects of type $A$.

If we introduce an object which is a sequence of objects of type $A$, then we say it is of type *seq A*.

Formally, given a set $A$ of objects, we define the set of sequences of objects of type $A$ as

$$seq\,A = \{f: \mathbb{N} \nrightarrow A \mid (\exists x: \mathbb{N} \bullet dom\,f = 1..x)\}.$$

That is, the set of sequences of objects of type $A$ is equal to the set of partial functions $f$ from the natural numbers to $A$ where there exists a number $x$ such that the domain $f$ is the set of numbers from 1 to $x$ inclusive. This means that a sequence can be any partial function from the set $1..x$ to the objects in $A$, where $x$ can be any natural number. Such a stipulation is necessary to ensure there are no 'holes' in the indexing domain. Thus, if $x$ were 10 then the sequence would contain 10 maplets each relating a number to an object in $A$.

**Example 7.2**

From the above, we can see that our sequence display notation is just an abbreviation for an equivalent set display:

- $\langle 6, 3, 2, 4 \rangle = \{1 \mapsto 6, 2 \mapsto 3, 3 \mapsto 2, 4 \mapsto 4\}$
- $\langle h, e, l, l, o \rangle = \{1 \mapsto h, 2 \mapsto e, 3 \mapsto l, 4 \mapsto l, 5 \mapsto o\}$
- $\langle \rangle = \{\} = \emptyset$

Note that the empty sequence $\langle \rangle$ is equivalent to the empty set.

**Example 7.3  WORD PROCESSOR continued...**

We can define the state space of the simple word processor using sequences. A word (type $WORD$) is a sequence of characters (type $CHAR$), and a document is a sequence of words.

$$WordProcessor \,\hat{=}\, [document : seq\,WORD]$$

There are two other types of sequence which are less commonly used, but may be useful in a specification. The first is the same as $seq\,A$ but does not include the empty sequence. It is called a **non-empty sequence**.

* $seq_1 A$
  ...denotes the set of non-empty sequences of objects of type $A$.

Thus, $seq_1 A = seq A \setminus \{\langle\rangle\}$.

We would use this notation when a model did not permit a sequence to be empty.

The next special sequence type is the set of **injective sequences** of objects. This is a useful idea since it represents the set of sequences which contain no repetitions—something which may be desirable in a specification. For example, an injective sequence could model a dinner queue; the first person in the sequence is at the head of the queue. The injective property asserts that one person cannot appear in the queue more than once, a fact which is hopefully true!

* $iseq A$
  ...denotes the set of injective sequences of objects of type $A$.

Thus, for a sequence of set $A$

$$iseq A = seq A \cap (\mathbb{N} \rightarrowtail A)$$

where $\mathbb{N} \rightarrowtail A$ is the set of all partial injections from the set of natural numbers to the set $A$. Hence, the intersection of $seq A$ and $\mathbb{N} \rightarrowtail A$ is the set of precisely those sequences of $A$ which are also injections.

# Indexing

The information held in a sequence can be accessed in a special way. When we used sets, we could only really test for membership—an object was either a member of a set or it wasn't. A sequence gives us extra power because we can not only find out whether an object is in a sequence or not (by using the *ran* operator with $\in$), we can also find what the first, or second, or fifteenth member is. This is known as **indexing** a sequence.

* $s\ n$
  ...denotes the object which is the $n$th member of the sequence $s$, i.e. if $s$ contains the maplet $n \mapsto x$, then $s\ n$ evaluates to $x$.

Clearly, indexing is just a special version of function application.

### Example 7.4

Suppose we are given the following sequence:

$s = \langle h, e, l, l, o, t, h, e, r, e \rangle$

then if we wanted to see if a was in the sequence, we would perform a membership test as follows:

$a \in ran\, s$

which would evaluate to *false* because a is not a member of the sequence $s$.

If we wanted to know what the fifth element was then we would use the following expression:

$s\, 5$

which would evaluate to the object o.

## Length

The length of a sequence is the number of elements in the sequence, i.e. the number of maplets in the sequence function.

*     $\#s$
  ...denotes the length of the sequence $s$.

Note that the length of the sequence is just the same as the cardinality of the set $s$ of maplets.

### Example 7.5     WORD PROCESSOR continued...

A useful function available when using the word processor is the ability to count the number of words in a document. This can be modelled with the cardinality operator.

```
┌─── WordCount ─────────────────────────────────
│ ΞWordProcessor
│ words! : ℕ
│ ───────────────────────────────────────────
│ words! = #document
└───────────────────────────────────────────────
```

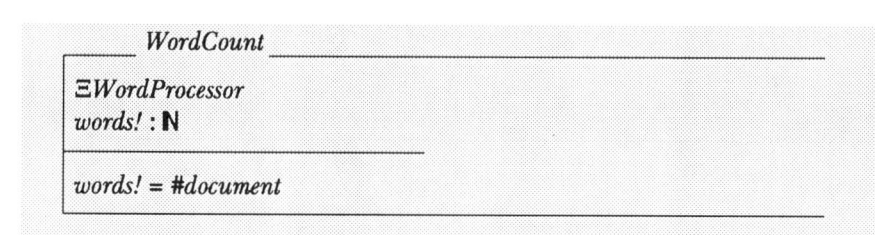

## 7.2 OPERATIONS ON SEQUENCES

Sequences of objects often crop up in real-world systems. Because they are so common, Z has a set of special purpose operators which can be used on them. These are in addition to the operations on relations and functions which can also be applied to sequences (remember, a sequence is simply a specific form of function which in turn is just a specific form of relation).

### Concatenation

Given two sequences $s$ and $t$, we can form a new sequence which contains all those elements of $s$ (in their correct order) followed by all the elements of $t$ (in their correct order). This process is called **concatenation**.

*     $s \frown t$
  ...denotes the sequence which is the concatenation of the sequence $s$ and the sequence $t$, i.e. the sequence of elements from $s$ followed by elements from $t$ in the correct order.

Note that the types of the two sequences must be identical.

### Example 7.6

The following are examples of concatenation:

- $\langle a, b, c \rangle \frown \langle d, e, f \rangle = \langle a, b, c, d, e, f \rangle$
- $\langle 3, 2, 1 \rangle \frown \langle 1, 2, 3 \rangle = \langle 3, 2, 1, 1, 2, 3 \rangle$
- $\{1 \mapsto h, 2 \mapsto e, 3 \mapsto l\} \frown \{1 \mapsto l, 2 \mapsto o\} = \{1 \mapsto h, 2 \mapsto e, 3 \mapsto l, 4 \mapsto l, 5 \mapsto o\}$

- ⟨⟩⁀⟨apple,pear,orange⟩ = ⟨apple,pear,orange⟩
- ⟨apple,pear,orange⟩⁀⟨⟩ = ⟨apple,pear,orange⟩

## Decomposition

A sequence has a **head** and a **tail**. The head of a sequence $s$ is simply the first element in the sequence, i.e. $s\,1$. The tail of a sequence $s$ is the sequence of all elements in $s$ other than the head of $s$, i.e. $\langle s\,2, s\,3, ..., s(\#s)\rangle$.

* *head s*
  ...denotes the object which is the head of the sequence $s$, i.e. the first element in the sequence $s$.

* *tail s*
  ...denotes the sequence which is the tail of the sequence $s$, i.e. the second element in $s$ followed by the third ... followed by the last element in $s$.

Thus, the head of a sequence and the tail of a sequence form the original sequence when concatenated

$s = \langle head\ s\rangle\frown tail\ s$

Note that we had to put *head s* in sequence braces because concatenation can only join two sequences together—it cannot join a non-sequence object to a sequence.

The operation *head s* is only defined when $s$ is a non-empty sequence—it doesn't make any sense to find out the first element of a set which has no elements! Similarly, *tail s* is not defined when $s = \langle\rangle$.

A sequence has a **last** element and a **front**. The last element of a sequence $s$ is $s(\#s)$. The front of a sequence $s$ is the sequence of all elements in $s$ other than the last, i.e. $\langle s\,1, s\,2, ..., s(\#s-1)\rangle$.

* *last s*
  ...denotes the object which is the last element in the sequence $s$.

* *front s*
  
  ...denotes the sequence which is the front of the sequence $s$, i.e. the sequence which contains all the element in $s$ in the correct order excluding the last element in $s$.

Thus, the front of a sequence and the last of a sequence form the original sequence when concatenated

$$s = front\, s \frown \langle last\, s \rangle$$

Note, as for *head* and *tail*, that *last s* and *front s* are only defined when $s$ is a non-empty sequence. Hence, the above expression only makes sense when $s \neq \langle \rangle$.

## Example 7.7

The following examples illustrate the use of *head*, *tail*, *front*, and *last* on sequences:

- $head\, \langle a, b, b, c, d, d \rangle = a$
- $head\, \langle a \rangle = a$
- $tail\, \langle a, b, b, c, d, d \rangle = \langle b, b, c, d, d \rangle$
- $tail\, \langle a \rangle = \langle \rangle$
- $front\, \langle a, b, b, c \rangle = \langle a, b, b \rangle$
- $front\, \langle a \rangle = \langle \rangle$
- $last\, \langle a, b, b, c, d \rangle = d$
- $last\, \langle b \rangle = b$

The relevance of the above four operations *head*, *tail*, *front*, and *last* is that they can be used to decompose a sequence. This is the opposite to concatenation. In concatenation we have a mechanism to add elements to a sequence, whereas the *decomposition* operators provide a mechanism for taking elements away from a sequence.

### Example 7.8    DINNER QUEUE

Earlier on we modelled a dinner queue as an injective sequence. The sequence could have the names of each person in the queue (each name being distinguishable from the other) mapped to by a number representing their position in the queue. Let us call such a sequence *queue*, with type *iseq NAME*.

The next person to be served in the queue is *head queue*, i.e. the first element of the sequence *queue*. When this person is served, they leave the queue and their name must be removed from the sequence. This can be achieved with the following operation:

*queue'* = *tail queue*

This means the person previously second in the queue is now first, and so on. The previous head of the queue has been removed from the sequence.

When a new person $x$ joins the end of the queue we use concatenation to add of an element to the sequence:

*queue'* = *queue* $\frown \langle x \rangle$

## Reversal

There is a special sequence operation which reverses the order of elements in a sequence.

* *rev s*
  ...denotes the sequence which is the reverse of $s$, i.e. it contains all the elements in $s$ but in exactly the opposite order.

### Example 7.9

The following provides an illustration of sequence reversal:

- *rev* $\langle 1, 2, 2, 3, 4, 5, 6 \rangle = \langle 6, 5, 4, 3, 2, 2, 1 \rangle$
- *rev* $\langle \rangle = \langle \rangle$
- *rev* (*rev* $\langle a, b, c, d \rangle$) = $\langle a, b, c, d \rangle$

# Filtering

One very useful sequence operation is known as **filtering**. This is the process of taking a sequence and 'filtering' out unwanted members of the sequence to produce a new sequence. It works like the restriction operators used on relations, by taking a sequence $s$ and a set $A$ which is of the same type as the range of the sequence (i.e. of the elements of the sequence), and forming a sequence which contains all those elements in $s$ which are members of $A$. The set $A$ filters out unwanted elements.

For example, if we had a sequence

$\langle a, b, a, d, c, b, c, e, a, e \rangle$

and filtered it using the set

$\{b, c, e\}$

then the resulting sequence would be $\langle b, c, b, c, e, e \rangle$.

*    $s \upharpoonright A$

    ...denotes the sequence $s$ filtered by the set $A$, i.e. the sequence which contains all those elements of $s$ which are members of $A$ in the same order as they appear in $s$.

## Example 7.10    DINNER QUEUE continued...

In the tradition of all good queues, no person is allowed to join the queue at any position other than its end! However, some bored person may decide to leave the queue and go to the nearest fast food restaurant. Say this person was $n$, then the following expression will remove that person's name from the sequence:

$queue' = queue \upharpoonright (NAME \setminus n)$

We have achieved this by filtering out $n$. The set $NAME \setminus n$ contains all possible names apart from $n$'s name, thus $n$ is excluded from the sequence $queue'$.

# 7.3 ADVANCED OPERATIONS ON SEQUENCES

## Distributed Concatenation

When we have a sequence of sequences, e.g. an object $ss$ of type $seq\ (seq\ A)$ for some object $A$, then we can perform an operation known as **distributed concatenation** on $ss$. What this does is result in a sequence which contains all the elements in $ss(1)$ followed by all the elements in $ss(2)$ and so on.

### Example 7.11

Suppose we have the following sequence of sequences:

$\langle\langle T, h, i, s\rangle, \langle i, s\rangle, \langle a\rangle, \langle s, e, q, u, e, n, c, e\rangle\rangle$

i.e. a sequence which has four elements, each of which are sequences themselves.

The distributed concatenation of the sequence is the following sequence:

$\langle T, h, i, s, i, s, a, s, e, q, u, e, n, c, e\rangle$

i.e. a sequence which has fifteen members.

* $\frown/ss$
    ...denotes the sequence which is the distributed concatenation of the sequence of sequences $ss$, i.e. the sequence which contains all the sequences in $ss$ concatenated with each other one after another.

An example of the use of distributed concatenation is when we have a sequence of sequences to model a two-dimensional table-like structure, and wish to know how many elements there are in the table.

### Example 7.12  WORD PROCESSOR continued...

Just like a *WordCount* operation, the word processor may have a character counting operation:

---
**CharCount**
---
$\Xi WordProcessor$
$chars!: \mathbb{N}$
---
$chars! = \#(\frown /document)$
---

Distributed concatenation is used to form a large sequence consisting of all characters in all the words in the document.

## Disjointness

If we have a sequence of sets, then the sequence is said to be **disjoint** when the intersection of all the sets is the empty set, i.e. if no two sets in the sequence have any members in common.

### Example 7.13

The following is an example of a disjoint sequence of sets:

$\langle \{a, b, e\}, \{c, d, f\}, \{g\}, \{h, l, m\}, \{i, j\} \rangle$

No two sets within the sequence have members in common. However, the following sequence is *not* disjoint:

$\langle \{1, 2, 3\}, \{4, 5, 6\}, \{7\}, \{8, 3, 9\}, \{10, 11\} \rangle$

The sequence is not disjoint because the intersection of the first and fourth elements of the sequence is {3}, i.e. they have a member in common.

Thus, disjointness is a possible property of a sequence (or an indexed function of sets). The following predicate can be used to test for disjointness:

✱   *disjoint s*
...denotes the predicate which is true precisely when no two sets indexed by *s* have members in common.

Note that *s* need not be a strict sequence. All that is required is that *s* is a function from an index to a set of objects, e.g. if the objects were of type *A* and the index set was *I*, then the type of the function must be

$\mathbf{P}(I \twoheadrightarrow \mathbf{P}A)$

A sequence has the natural numbers as its index set, with its domain as all values between 1 and the number of elements in the sequence inclusive.

The following is a more formal definition of the disjointness property of a sequence *s*:

$disjoint\, s = (\forall\, x, y : dom\, s \mid x \neq y \bullet s\, x \cap s\, y = \emptyset)$

That is, for a set *s* to be disjoint the following must be true: "for every *x* and *y* which are different indices of *s*, the intersection of the set *s x* and the set *s y* is empty".

## Partitions

There may be some cases where we want to take a collection of objects and **partition** them into several disjoint smaller collections. The idea is analogous to cutting up a cake into several slices. The point being that when the slices are joined together, they form the complete cake again.

The notion of a partition can be modelled in Z. One way is to have a set of sets—each member set models a slice of the cake, and the union of all the member sets forms the cake as a whole. This is a reasonable way to model a partitioned entity, but we will consider a sequence of sets. This allows us to index each slice which is a useful aid to flexible modelling.

A partitioned sequence of sets of type $\mathbf{P}B$ is a sequence which is disjoint and whose range contains sets whose combined union forms the set of all the members of the set *A*, where *A* is of type $\mathbf{P}B$.

**\*** *s partition A*
...denotes the predicate which is true precisely when the sequence of sets *s* partitions the set $A$, i.e. when *disjoint s* is true and the union of each member set in the range of *s* is equal to the set $A$.

As for the *disjoint* predicate, *s* above does not have to be a sequence of sets; any indexed function of sets will do.

## Example 7.14

Suppose we were managing a consultancy. At any one moment in time there are many projects active. We must assign each consultant to a team which concentrates on one project. No single consultant can be on more than one project at the same time. This structure lends itself to being modelled as a sequence *project* of sets, representing a list of project sets. The names of the consultants are members of the sets.

We can assert the following:

*project partition consultants*

This asserts that the sequence *project* partitions the set of all consultants in the firm.

# 7.4 SUMMARY

Sequences are very useful in modelling information systems. They provide a convenient mechanism for indexing data—each item of data is mapped to by a different number. This closely reflects the way that data is stored in computer memory.

The most important property of a sequence is the fact that it is an ordered list of objects. This means that we may have some items appearing more than once in a list, and that we know the precise location of each data item. This is a property which can be exploited to model quite complicated data structures.

# CHAPTER EIGHT

# Bags

The previous chapters have introduced some useful mathematical data types along with their own special notation in Z. These data types tend to crop up often when modelling systems. In this chapter we introduce the concept of a **bag**.

A bag is another specialised collection of objects. It consists of a list of objects along with the frequency of occurrence of each object. In other words, it is a data type that specialises in dealing with repeated objects; as such it is a cross between a set and a sequence in expressive power[1]. It is useful if we want to easily keep track of the number of repetitions of a particular object.

For example, if we had a coin counting machine, we might like a report produced formatted as a list of each coin denomination followed by the number of coins counted of that denomination, e.g. 1p×47, 2p×79, 5p×53 etc...

---

[1] The term *expressive power* is used when describing the ability of a notation to mimic real-world information. A notation with a high degree of expressive power is said to be able to capture complex real-world concepts allowing their easy and meaningful manipulation.

*Definition*

A bag is a partial function from a set of objects to the positive natural numbers, i.e. a bag $b$ of objects of type $A$ has the type $A \twoheadrightarrow \mathbf{N}_1$, where $\mathbf{N}_1$ denotes the set of all natural numbers greater than or equal to one.

*Notation*

There is a special notation for displaying bags of elements. We enclose the objects within the square brackets " [ " and " ] ", and display each object as many times as it appears in the bag.

## Example 8.1

The set $\{a \mapsto 3, b \mapsto 2, c \mapsto 1, d \mapsto 1\}$, which represents the bag containing three $a$'s, two $b$'s, a $c$, and a $d$, can be written as

$$[a, a, a, b, b, c, d].$$

Thus, we can display bags using either methods. Note that a bag display can have repeated members. Set display does not have repeated members. The following illustrates this:

I   $[1, 2, 3] \neq [1, 2, 2, 3]$

II  $\{1, 2, 3\} = \{1, 2, 2, 3\}$

The sets in II are equivalent because they both represent a set which has the distinct members 1, 2, and 3, hence being of cardinality three (although the LHS of II is strictly the correct way of displaying the set in Z). However, the LHS of I is the set

$$\{1 \mapsto 1, 2 \mapsto 1, 3 \mapsto 1\}$$

which is different to the RHS of I which is the set

$$\{1 \mapsto 1, 2 \mapsto 2, 3 \mapsto 1\}.$$

The empty bag is []. It represents a bag of no elements, i.e. a bag whose domain is the empty set. Thus,

$$[\,] \Leftrightarrow \emptyset$$

i.e. the empty bag is equivalent to the empty set.

## Bag Types

It has already been stated that a bag of objects from $A$ is of type $A \nrightarrow \mathbb{N}_1$. However, there is a special notation which is equivalent to this bag type:

*     $bag\ A$
      ...denotes the set of all bags of objects of type $A$.

If we declare an object $x$ as a bag of objects of type $A$, its type would be $bag\ A$.

# 8.1 FUNDAMENTAL CONCEPTS

## Bag Membership

We can test to see if a particular object is in a bag using the **bag membership** operator.

*     $x\ in\ b$
      ...denotes the predicate which is true precisely when the object $x$ is a member of the bag $b$.

The *in* operator works in exactly the same way as $\in$ does for set membership. In fact, *in* is just an abbreviation for $\in$:

$$x\ in\ b \Leftrightarrow x \in dom\ b$$

because the domain of the bag $b$ is the set of objects in $b$.

## Counting

One of the major benefits of modelling structures with bags is that they keep a count of the number of occurrences of each object within them.

There is a special bag operation which, given an object *x* and a bag *b*, evaluates to the number of *x*'s in *b*:

*     *count b x*
    ...denotes the natural number which represents the number of times the object *x* appears in the bag *b*.

## Example 8.2

Given the bag

$$b = [a, a, a, b, c, c, c, c, d]$$

then the following are true:

- *count b* a = 3
- *count b* d = 1
- *count b* e = 0

## Creating Bags from Sequences

Bags can be used hand-in-hand with sequences. A sequence is an ordered list of objects with the possibility of repetition—an element can occur any number of times in the list. We can take such a sequence and turn it into a bag of the objects in the sequence.

*     *items s*
    ...denotes the bag which contains exactly the same elements as the sequence *s*, i.e. the bag of objects in the sequence *s*.

## Example 8.3

If we had a sequence *todo* which modelled an "in tray" on a clerks desk, then the head of *todo* would be the next job to be processed by the clerk. There are five broad categories of job which the clerk may be faced with: an invoice, a memo, an order, a job application, and a report. They are members of the following set:

*job* = {invoice,memo,order,application,report}

Sometimes the clerk wants to know how many of each job he has yet to do. We can find this out by creating a bag from *todo*:

*jobbag* = *items todo*

To find out how many invoices are to be processed, we can perform the following operation:

*count jobbag* invoice

which evaluates to exactly the number of invoices in the "in tray". Finding such a value could have been achieved without creating a bag as follows:

#{$x : dom\ todo\ |\ todo(x)$ = invoice}

It is clear that creating a bag is far more intuitive. The fact that we are interested in the number of invoices is far more obvious when we use the *count* operator.

# 8.2 OPERATIONS ON BAGS

There is only one built-in special purpose operation that can be performed on bags to manipulate them. However, as bags are just partial functions, all the relation and function operations can be performed on them.

### Bag Union

Given two bags of the same type, we can form a new bag which joins the two together. Any object which appears in both bags has its number of occurrences in each added in the new bag. This is called **bag union**.

✱    $b \uplus c$
        ...denotes the bag which is the union of the bag $b$ and the bag $c$, i.e. the bag which contains each element in $b$ and $c$ as many times as it appears in both $b$ and $c$ added together.

If we have two bags, $b$ and $c$ of the same type, then the following observations are true:

- $dom\ (b \uplus c) = dom\ b \cup dom\ c$
- $count\ (b \uplus c)\ x = count\ b\ x + count\ c\ x$

i.e. the bag union of two bags contains all the elements in each bag, and the count of each element in the bag union is the sum of the count in each bag separately.

**Example 8.4**

The following are examples of bag union:

- $[a,b,c,c] \uplus [a,c,d] = [a,a,b,c,c,c,d]$
- $\{a \mapsto 6, b \mapsto 12, c \mapsto 4\} \uplus \{a \mapsto 1, d \mapsto 1\} = \{a \mapsto 7, b \mapsto 12, c \mapsto 4, d \mapsto 1\}$
- $[a,a,a,b] \uplus [\ ] = [a,a,a,b]$

### Bag Summary

Bags provide us with an easy method of keeping a count of the number of objects in a structure. The special bag operators allow us to utilise this power in an easy and obvious way. Using bags in this way can help the reader understand a specification better.

## 8.3 THE Z DATA TYPE PYRAMID

In this book we have introduced *sets* and shown how they underpin the standard Z mathematical data types like relations, functions, and sequences. Indeed, set theory along with *logic* provide the foundations for the entire Z Notation, and the subsequent specifications written in Z.

Along with sets we introduced the *cartesian product*—a way of linking sets together. Through these links we can create *ordered-pairs*. By following a standard set of conventions we can say that the ordered pair $(x, y)$ denoted a relationship between *x* and *y*. This gives a new general-purpose data type called a *relation*. A certain restricted form of relation found often in the real world is the *function*, which can model many-to-one and one-to-one relationships. Using functions it is possible to model an ordered *sequence* of objects, and a *bag* of objects which keeps track of the number of occurrences of objects in a collection.

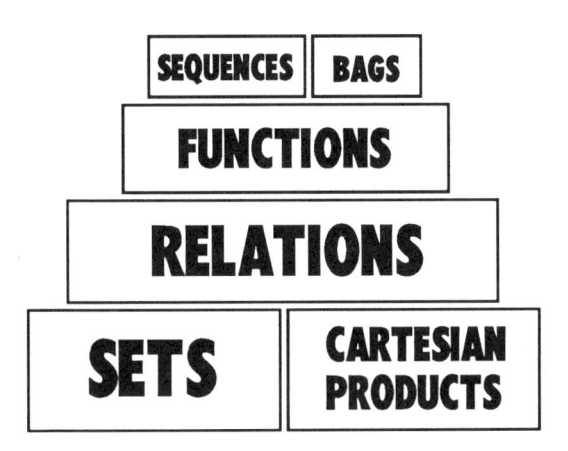

**Figure 8.1**

Figure 8.1 depicts the data type 'pyramid' which is evident from the dependencies of the data types. These are the standard Z data types. From these it is possible to devise other data types which may particularly suit a specific model.

## Summary Of Data Type Characteristics

The following table provides a brief summary of the characteristics of the Z data types:

| DATA TYPE | CHARACTERISTICS |
|---|---|
| *Sets* | Unordered. No Duplicates. Unstructured. Very Flexible. |
| *Relations* | Relationships: m-n, 1-1, 1-n, n-1. Flexible. Many operators. |
| *Functions* | Special Relationships: n-1, 1-1. Useful operators. |
| *Sequences* | Ordered. Duplicates. Specialised. Useful operators. |
| *Bags* | Counts Duplicates. Very Specialised. |

# CHAPTER NINE

# Advanced Z

This chapter contains some of the more advanced topics in Z. It is aimed at those who are interested in delving deeper into Z, but is not essential reading for an introductory book. Some of the notation may crop up in future exposure to Z, in which case it will be necessary to read the relevant sections herewithin (for example, this may be necessary when reading over a Z specification document which contains previously unseen notation).

The concepts explained or outlined in this chapter are not strictly necessary for an understanding of the most common uses of Z, nor are they strictly necessary for an appreciation of Z as a powerful specification language which can be used to specify real world systems.

For those who are interested in reading on, the chapter is divided into three sections. The first two deal with advanced notation: one dealing with schemas in the same degree of detail as the other subjects covered previously in the book, and the other briefly introducing several other advanced topics in Z. Not every advanced feature of Z is covered. Some of the more obscure ones are left to other more advanced Z books to be studied if necessary. The second section also lists the subjects not covered at all in this book. The last section deals with mathematical proof. It tries to provide a gentle introduction to what many perceive as a challenging, highly theoretical subject. It is hoped that the coverage given will show that the ideas behind proofs are relatively easy to comprehend.

## 9.1 ADVANCED USES OF SCHEMAS

Schemas have state. To recap, a schema $S$ introduces objects in its signature, and constraints on the values that these objects may take in its predicate part. The state of $S$ at any instance is the value of each of the objects in its signature. The state space of $S$ is the set of all possible legal states $S$ can be in, i.e. it is the set of all possible combination of values assignable to the objects in the signature satisfying the specified constraints of $S$.

### Bindings

Suppose we have a schema $S$ with the following identifiers in its signature:

$x : A$

$y : B$

Then a **binding** of $S$ is an instance of $x$ and $y$ in the current environment. $S$ by itself represents the set of all bindings which satisfy the predicate part of $S$. A binding can be a particular state of a schema.

*     $\theta S$
    ...denotes a binding of the schema $S$.

$\theta S$ represents any binding of $S$. The theta prefix is called the **binding formation** operator. $\theta S$ can be a particular state of a schema $S$.

### Example 9.1
For example, let us consider the following schema:

```
┌─── Uneq ──────────────────────
│ x, y : {1, 2, 3}
│──────────────────────────────
│ x ≠ y
└──────────────────────────────
```

Now, this definition describes a schema *Uneq* by defining all the possible legal states it can be in—all its possible legal bindings. It does this by stating that *Uneq* consists of two objects, that each object can take a value of 1, 2, or 3, and that the value of the two objects must differ. Thus θS could be represented by a binding where

$$x = 2 \land y = 1.$$

## Schema Types

When we say an object $f$ is of type $A \twoheadrightarrow B$ we are saying that $f$ is a set of maplets such that the domain is a subset of $A$ and the range is a subset of $B$. Z also allows objects to have **schema types**. For example, if we have a schema $S$, then we can say an object $x$ is of type $S$. This means that $x$ can be any instance of $S$—it can take on the value of any binding θS in $S$ which satisfies the constraint in $S$.

### Example 9.2

Consider Example 9.1. Suppose we have an object $z$ of type *Uneq*, then $z$ is a compound object formed from two objects each of type {1, 2, 3} such that the objects are unequal. The object *inherits* the property of the schema.

We need some new notation for representing schema types:

* ⦇ $x : A; y : B$ ⦈
    ...denotes the schema type of a schema with objects $x : A$ and $y : B$ in its signature, i.e. the set of all bindings of a schema including those bindings that do not satisfy the predicate part.

In practice we do not use this notation often because if we wish to introduce an object *x* which is an instance of a schema *S*, then we declare it by using the schema name as the type name.

Schema types are a very powerful feature of Z. They are closely related to the concept of *abstract data types*. This is when we model an entity by taking the constituent data components and the legal operations that can be performed on the data, and package them together as a single object which can be named.

The benefits to be gained from abstract data typing include:

- *Abstraction*: being able to refer to a complex idea depending on many different data entities using a single name.
- *Consistency*: by using the same objects and operations for the same concepts throughout a specification.
- *Indivisibility*: a schema can be thought of as a single entity which can be manipulated using only the given allowed schema operations. As such, the data within a schema are not treated as elementary objects but rather as a set of interrelated objects which only taken together represent a meaningful real-world concept.
- *Single Reference Point*: the operations are clearly, centrally defined so the legal operations on the data can be swiftly referenced.

## Example 9.3   DATES

There are many programs which deal with dates, e.g. appointment diary programs, accounting programs, scheduling programs etc. The date used in a particular system may be composed of four objects; a date-in-the-month, a month, a year, and from these one can compute the day in the week.

We can benefit from abstraction here by gathering the four components into one entity called a *Date*:

┌─── *Date* ─────────────────────────────────
│ *day* : {mon, tue, wed, thu, fri, sat, sun}
│ *date* : 1..31
│ *month* : 1..12
│ *year* : **N**
│ ─────────────────────────
│ $date \leq days\_in(month, year) \wedge$
│ $day = day\_for(date, month, year)$
└────────────────────────────────────────────

The predicate part asserts that the date-in-the-month should not exceed the number of days in that month by using a function called *days_in* which maps each (month, year) pair to the number of days in the month (we need to know the year because there are 29 days in February on leap years). Another function *day_for* maps each (date, month, year) 3-tuple to the day in the month in which that date falls on.

*Date* is an abstract data type. From now on, whenever we wish to refer to a date, or perform some operation on a date, we can introduce an object of type *Date*.

## Selection

When we have an object which is a schema type, we may wish to refer to the value of its component objects. We can do this by using **selection**. This takes a binding of a schema and one of the schema's components, and evaluates to the value of that component in the given binding.

∗    $z.x$
    ...denotes the value of the object $x$ in the binding $z$.

Thus, if we had the object $z = \{\!\!\{ x : A; y : B \}\!\!\}$, we could determine the value of $x$ with the expression $z.x$.

## Example 9.4  DATES continued...

Suppose we were modelling a computerised life assurance company. The state space of the system is called *Assurance* (the details of which we are not interested in here).

The company has many policies taken out by clients each year which expire up to 25 years later. For each policy we wish to know the policy number (of type *PN*), the policyholder's name (of type *NAME*), the date the policy was taken out (of type *Date*), the period over which the policy will be active in years, and other financial & personal details (of type *INFO*). Thus, we can define a policy data type as:

―――― *Policy* ――――――――――――――――――――
$num : PN$
$name : NAME$
$takenout : Date$
$period : 1..25$
$details : INFO$
―――――――――――――――――――――――――

The process of paying out a due policy starts three months before the expiry date so that the financial calculations can be made, and the client informed. So, every month the computer must present a list of policies that will expire three months later. We can model this using the schema *Expires* which assumes the existence of a global variable called *projected* of type *Date* which holds the current date plus three months, and that the state space *Assurance* has a database of policies taken out modelled as a set called *policies* of type **P***Policy*:

―――― *Expires* ――――――――――――――――――
$\Xi Assurance$
$list! = \mathbf{P}PN$
―――――――――――――――
$list! = \{x : PN \mid (\exists y : Policy \bullet y \in policies \land x = y.num \land$
$\qquad\qquad y.period + y.takenout.year = projected.year \land$
$\qquad\qquad y.takenout.month = projected.month)\}$
―――――――――――――――――――――――――

This schema outputs the set *list!* of policy numbers for the policies that will expire in three months time. The set is defined as all those policy numbers such that the policies with those numbers were taken out exactly as many years ago as the duration of the policy and in the same month as the current date projected three months in the future.

We used selection many times. The predicate $x = y.num$ is true precisely when the policy $y$ we are considering is numbered $x$, i.e. the *num* component of the binding $y$ of *Policy* is equal to $x$.

The expression $y.takenout.year$ evaluates to the year in which policy $y$ was taken out. This uses selection twice; $y.takenout$ represents the date in which the policy was taken out, and the ".*year*" suffix evaluates to the year component of that date.

The above example really does benefit from schema types. We were able to define an abstract data type called a *Date* and another one called a *Policy* which used *Date*, and we were then easily able to refer to "something which is a policy" and "something which is a date" in an intuitive manner. We captured the concepts of date-like and policy-like objects.

## 9.2 MISCELLANEOUS Z NOTATION

This section aims to give a brief discussion on some of the more advanced topics and notation in Z which do not need to be known at this level of study. However, those wishing to be able to write rigorous Z Specifications should read some of the literature mentioned in Chapter One[1].

### Generics

So far, when illustrating a new piece of Z Notation, we have rarely given the proper formal definition of the notation—we have used fairly informal

---

[1] *Software Engineering Mathematics* and *The Z Notation: A Reference Manual* are two good books to start-off with.

descriptions plus examples. When we did give a more formal definition, we relied on specific examples. For example, our 'formal' definition of a sequence was

$$seq A = \{f: \mathbb{N} \nrightarrow A \mid (\exists x: \mathbb{N} \bullet dom f = 1..x)\}.$$

This definition only holds if we want a sequence of objects of type $A$. Of course, we generally want a free choice of the types we can use—we don't want to stick with just $A$.

Z overcomes this problem with special constructs known as **generic schemas**. These are schemas which contain local basic type declarations known as **formal generic parameters**. Later, these formal parameters can be assigned with the actual types to be used—the **actual generic parameters**.

Consider the following example of a generic schema:

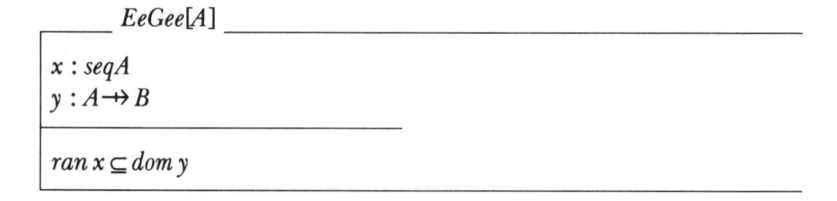

```
┌─────── EeGee[A] ─────────────────────────
│ x : seqA
│ y : A ↛ B
│ ─────────────────────
│ ran x ⊆ dom y
└──────────────────────────────────────────
```

This schema introduces a sequence of objects of type $A$, where $A$ is a local basic (generic) type (i.e. a type which is not yet specified), and a partial function from this $A$ to a known type $B$ (defined elsewhere). $A$ could be any set, e.g. $\mathbb{N}$, or the set of all names, or a set of coin denominations etc.

$A$ is really just a placemarker. When we refer to the schema, we must provide the actual parameter, i.e. we must state precisely what $A$ is—what to replace $A$ by. For example, suppose we want $A$ to be the set of names (called *NAME*), then we would refer to:

    *Eegee*[*NAME*]

which is equivalent to the schema:

$$\begin{array}{|l|} \hline x : seqNAME \\ y : NAME \rightarrowtail B \\ \hline ran\ x \subseteq dom\ y \\ \hline \end{array}$$

There is another useful form of generic construct that can be used in Z; they are called **generic constants**. These introduce constant-valued global objects whose type we do not yet specify. Thus, they are like axiomatic descriptions (because they are global) with generic parameters.

The objects are declared and defined in a frame like a schema. However, generic constant definition frames never have names; instead they have a declaration of the formal parameters, and the top line is a double line to distinguish them from schemas. The following defines two generic constants built into the Z Notation, and are useful when dealing with ordered pairs:

$$\begin{array}{|l|} \hline [A, B] \\ \hline first : A \times B \rightarrow A \\ second : A \times B \rightarrow B \\ \hline (\forall\ x : A; y : B \bullet first(x, y) = x \land second(x, y) = y) \\ \hline \end{array}$$

Here we are saying that *first* is a total function from all ordered pairs of type $A \times B$ to the set $A$ such that *first* applied to a pair equals the first element of that pair. A similar explanation holds for the function *second*. These functions can be used in any Z Specification. Note the generic nature of the operators: they work on any cartesian product of two sets.

## Example 9.5

If we had the object:

    *name* = (John, Smith)

then we could find the surname of the person with the expression

    *second name*

which would evaluate to "Smith".

Another example of using generic constants is to introduce convenient abbreviations into specifications. These are called **abbreviation definitions**, and are a special form of generic constant. They use a double equals sign to define an identifier to equal some expression. For example, we can define the set of strictly positive integers as follows:

$\mathbb{N}_1 == \mathbb{N} \setminus \{0\}$

This says that the notation $\mathbb{N}_1$ is an abbreviation for the set $\mathbb{N} \setminus \{0\}$.

## Lambda Expressions

The Z Notation allows the use of **lambda expressions**. These are expressions which describe functions, and are extensively used in the *lambda calculus*[2] as a mechanism for providing formal semantics to computing problems.

The notation for lambda expressions in Z is as follows:

\*   $(\lambda x : A \bullet E)$
    ...denotes the function whose argument is of type $A$ and whose result is given by the expression $E$.

The expression in brackets describes a set, just as set comprehension does, but in this case the set is a function. The argument of the function is given by the general shape of the declaration in the expression, i.e. the objects declared after the lambda symbol and before the • separator. The resulting value of an argument applied to the lambda function is described by the expression after the separator.

---

[2] The lambda calculus is a mathematical formalism developed by Alonzo Church. Whilst originally intended to be used to study the behaviour of functions and function application, the calculus has also been used to help define the semantics of programming languages. A. Church, *The Calculi of Lambda-Conversion* (Princeton University Press, 1941).

## Example 9.6

Suppose we wanted to describe a function which returns the cube of an integer argument. We could define the following lambda expression to achieve just that:

$$cube = (\lambda x : \mathbb{N} \bullet x^3)$$

This says that when we apply the function *cube* to a natural number $n$, then the result will be $n^3$.

For example,

$$(\lambda x : \mathbb{N} \bullet x^3)\, 3 = 27$$

## Example 9.7

The *count* operator can be defined using a lambda expression. If we have a bag $b$ of type $bag A$ then $b$ is a partial function from the set $A$ to the set of positive natural numbers. If we knew that $b$ contained occurrences of the object $x$ then we could simply count the number of occurrences using function application:

$$b\, x$$

However, we may not be sure that another object $y$ is in the bag. If we use function application 'blindly', the result may be undefined and hence the system could be in an illegal state. We can't accept this, so we get around the problem by defining a total function *count b* which when applied to any member of $A$ gives a defined result (zero if the object is not in $b$):

$$count\, b = (\lambda x : A \bullet 0) \oplus b$$

The lambda expression describes a total function which maps every member of $A$ to zero. That function is then overridden with the bag partial function $b$. Thus, the resulting function maps each object in $b$ to the number of times it occurs in $b$, and all other members of $A$ not in $b$ to zero.

### More Schema Expressions

There are some schema operators other than the five we considered in Chapter Four. We will briefly discuss one of them in this sub-section. It is particularly useful when we are dealing with systems which are inherently sequential in nature, i.e. one process occurs before the next process, and the order of the processes is important.

If we have two schemas $S$ and $R$, then we can combine them to form a schema which does both their jobs. However, for this to happen, the outputs of $S$ must match the inputs of $R$, and the dashed variables in $S$ must match undashed versions of the variables in $R$. The combined schema is called the **schema composition** of $S$ and $R$.

✻     $S \fatsemi R$
    ...denotes the schema composition of $S$ and $R$.

### Z Notation Not Covered

The following is a list of some of the notation not covered in this book:

- Schema Operators: $\exists, \exists_1, \forall, \setminus, \upharpoonright, pre$
- Free Types
- Generalised Union and Intersection

## 9.3 AN INTRODUCTION TO PROOFS IN Z

This section takes a look at Z from a more theoretical viewpoint. It shows how the formal, rigorous foundations of Z work to enable us to prove useful things about our specifications with a degree of certainty which simply cannot be achieved using conventional development methodologies[3]. The

---

[3]    The system of inference rules for deriving theorems about Z specifications is still evolving. Hence, this section serves merely to introduce the important concepts involved.

subject matter may need to be read over a few times for all the concepts to sink in.

## Z : A Formal Notation

The **Z Language** is a *formal language*. It is composed of an **alphabet** (the symbols used in Z Specifications), and a set of **syntax rules**; the latter specify how the symbols may be combined in various ways to form 'legal' Z sentences[4] (these sentences, or concatenations of Z symbols, are called **well-formed formulae**). This book has described much of the Z Language.

The Z Notation contains the two vital ingredients of a formal notation: a formal language (the Z Language) and a set of **semantic rules** which allow us to interpret the well-formed formulae of the language. Without semantic rules we would not be able to give meaning to Z Specifications; we would not be able to take a well-formed formula in Z (which is simply a collection of funny looking symbols) and understand what it was saying about our computer system.

Semantic rules are hard to specify rigorously; consequently there is no universally accepted standard for doing so. The problem is exacerbated by the complexity of the Z Language. To overcome this problem, this book has resorted to using English explanations of the meanings of the various well-formed formulae in Z. This is not an ideal state of affairs; however it is hoped that using familiar mathematical concepts and explaining new ones clearly, reduces the scope for misinterpretation[5].

## Deduction and Proof in Z

Suppose we are building a Z Specification. At each stage, when we add each new Z sentence, we should do three things:

---

[4] The official Z Notation formal language is contained in Spivey's *The Z Notation: A Reference Manual*. It shows the full alphabet of Z, and describes its syntax rules using a meta-notation known as BNF.

[5] Spivey's *Understanding Z* contains a more rigorous look at Z. However, it should be noted that the approach taken is very advanced indeed, and most will find it hard going.

- Ensure the sentence is a well-formed formula by checking that all the symbols are in the Z alphabet, and that the symbols go together in accordance with the Z syntax rules.
- Show that the sentence is makes sense. So, not only is it syntactically correct, but it also has an interpretation in Z (so it evaluates to *true* in legal states).
- Check that the new sentence does not contradict any of the existing sentences when added to the specification.

The last two obligations each require a **proof**. In the context of Z, these show unequivocally that the sentence added is true, or that adding it to the true specification will not make the specification false[6].

In order to carry out a proof formally we need a *deductive apparatus* which augments the Z Language. This creates a *formal system* that allows us to deduce the new Z sentences from existing Z sentences in the specification and other sentences which are taken to be true. This formal system makes no reference to the semantics of Z. We simply take one string of symbols and deduce another string from it, hence it is a purely syntactic device which is independent of any interpretation we may wish to put on the language. In this way, if we deduce one sentence from another, it cannot be denied since we took a purely objective approach, and did not deal at all with the more subjective semantics of the sentence.

The deductive apparatus in Z and any other formal system is composed of:

- A collection of well-formed formulae (a subset of the language) which need not be proven. These are the **axioms** of the formal system, and are the building blocks used to derive other often more complicated sentences.
- A set of **inference rules** which allow us to infer one well-formed formula as an immediate consequence of other well-formed formulae.

---

[6] Also, by ensuring that a specification evaluates to true under all possible real-world scenarios we can show that it is complete—that all the states in which the specification can be in represent all the states the computer system could be in.

A well-formed formula which has been inferred using one or more inference step is called a **theorem**.

Now, the power of our formal system is that if we *assume* our Z axioms are true, and that our inference rules are well-founded, then any Z theorem is also true[7].

Formally, a Z proof is a sequence of Z sentences which are either axioms, theorems, or immediate consequences of one or more of the preceding Z sentences.

### Example 9.8

Suppose the following were axioms in Z:

- $a$
- $b$

and that the following are inference rules in Z:

- $a \wedge b \Rightarrow c$ (Rule 1)
- $c \Rightarrow d$ (Rule 2)
- $d \Rightarrow e$ (Rule 3)

Now, suppose we wanted to add a well-formed formula $d$ to our specification. The following is a proof of $d$:

| INFERENCE STEP | EXPLANATION |
|---|---|
| $a$ | Axiom |
| $b$ | Axiom |
| $a \wedge b$ | Consequence of the previous two steps |
| $c$ | By applying rule 1 to the previous step |
| $d$ | By applying rule 2 to the previous step |

---

[7] A formal system is said to be *consistent* if every theorem of the system is true. A system is *complete* if every true statement can be proven. Z is consistent but not complete. Hence, all things we prove will be true, but we won't be able to prove everything which we know to be true.

In other words, we assume that $a$ and $b$ are true, therefore $c$ is true according to inference rule 1, and hence $d$ is true (it is the immediate consequence of $c$ using rule 2). $d$ is now a theorem, and hence we can prove that $e$ is true.

In summary, if we take the Z alphabet, we can produce an infinite number of possible sentences. However, the syntax rules assert that only a subset of these sentences are legal Z sentences (well-formed formulae of the Z Language). Furthermore, only a subset of the well-formed formulae are theorems in a particular Z specification.

Proofs ensure that a Z Specification stays within bounds that allow us to spot errors at an early stage (when they are relatively cheap to correct) and be confident that what we have is consistent and correct. Of course, a specification may be 'correct' but still not specify what you want—it could be specifying what you don't want in a correct way! *This is the ultimate limitation of any specification language.*

THE END

# APPENDIX A

# Case Study

This appendix considers the Z Notation covered in the nine chapters of this book with the aim of specifying a fictional, yet hopefully realistic case study. It is not intended to be too complex—the aim is to provide a reasonable demonstration of Z, and no attempt is made to show how Z can be used in very complex scenarios; that is left to more advanced books on the subject. At this point, the reader who is interested may wish to refer to *Specification Case Studies* (see Chapter One) for interesting, but advanced real-life examples.

We will outline a large section of the system to be specified and then proceed with the Z specification for that section alone. Comments which are not necessary for the specification but are included for explanatory purposes are shown in a narrow sans-serif font e.g. "This is a comment and not part of the Z Specification."

# A SIMPLE BANKING SYSTEM

A computer consultancy firm called Haque Eurse Limited, is approached by a small bank called Quapp, Tôle & East plc (QTE), to investigate the computerisation of its current account facilities.

The bank has around 100,000 current accounts. Each customer can have several accounts. Each account is registered under the name of a single person. Statements, cheque books, and customer information are sent to the customer's registered home address.

No customer is allowed to have a negative balance unless an overdraft is approved by the bank manager. Unapproved indebtedness is charged interest at an amount equal to the Default Overdraft Surcharge (DOS) above the Standard Bank Rate (SBR). The default Customer Credit Limit (CCL) is £0, but an approved overdraft can increase this e.g. a CCL of £250 means the customer can have a balance of -£250 and not get charged the DOS.

The bank employs a "free banking" system of no charges for transactions.

Each customer has two balances: an account balance and an available balance. The available balance is the amount of money the customer can withdraw at any moment in time (this may be less than the expected account balance). Cash deposited becomes immediately available and thus the account balance and available balance increase by the same amount. When a cheque is deposited into an account, the account balance is increased, but the available balance does not increase until the cheque is cleared approximately three working days later. The bank cannot accept any other form of deposit.

Withdrawals cannot cause the customer's available balance to fall below the allowance dictated by CCL.

A log is kept of all the transactions carried out in the bank for each customer. Records are only kept for ten years.

The bank expresses an interest in providing seven days-a-week service using Automated Teller Machines (ATM) placed in strategic locations outside some branches and in some shopping centres.

The firm of consultants come up with a proposal, from which the following is an extract:

## 1 HARDWARE

The following hardware configuration is recommended:

1.1 The bank should purchase a mainframe computer which will act as the central computer of the firm.

1.2 Each branch should have several intelligent terminals, including one for each cashier. These will be linked to the main computer by telephone lines and leased lines.

1.3 ATMs will be linked to the main computer by telephone lines. Customer's are required to enter a four-digit Personal Identification Number (PIN) in order to make a transaction at an ATM.

1.4 ...

## 2 CASHIER TERMINAL SOFTWARE

The software written for the system of cashier terminals should include the following functionality:

2.1 Open an account for a new customer.

2.2 Open a new account for an existing customer.

2.3 Close an account.

2.4 Order a statement.

2.5 Print a statement of balance.

2.6 Deposit cash or cheques into an account.

2.7 Withdraw of cash, or cheque, without reducing the balance below that allowed by the CCL.

2.8 Transfer funds between customer accounts.

2.9 ...

**3    ATM TERMINAL SOFTWARE**

The software written for the system of ATMs should include the following functionality:

3.1  Withdraw up to £100 in cash per day without reducing the balance below that allowed by the CCL.

3.2  Print a statement of balance.

3.3  Order a statement.

3.4  ...

**4    GENERAL**

4.1  The log is to be kept on-line (directly accessible) for the last twelve months of transactions for each customer. Previous transactions up to ten years will be archived and hence there will be a delay of 24-hours when retrieving this data.

4.2  An ATM card is issued to each customer who wishes to make use of the ATM facilities. The magnetic strip on the card contains the card number. The PIN number is held on the main computer. Each card is associated with precisely one account.

4.3  ...

The system is to be specified formally as a Z specification document.

*Z Specification of a Simple Banking System (extract)*

The information stored about a customer can be divided into four categories, each of which have a different degree of permanence and use:

1.   The customer's identification. This can be used when proof of identification is required, or when the customer needs to be contacted (e.g. when sending statements). It is likely that this information will not change much in the lifetime of the account.

2. The details held about the customer. This includes the CCL, the PIN number of the ATM card the customer owns (if at all), and other relevant information (such as credit worthiness etc.). Such data may change a few times a year.

3. The current account balance and available balance. These balances will be frequently referenced and altered.

4. The transaction log. All transactions are recorded within this log, and hence this changes very often.

The above four facets will be modelled with four schema types, each with a different number of components reflecting the different attributes contained by the facet.

The first facet has a schema type named *IDENT* and contains three components relating to the name, address, and telephone number of the customer:

$$
\begin{array}{|l}
\hline
\quad \textit{IDENT} \\
\hline
\textit{name} : NAME \\
\textit{address} : ADDR \\
\textit{phone} : PHONE \\
\hline
\end{array}
$$

The second facet will be modelled using the schema type *DETAIL* which has three components: the CCL, the PIN number for the ATM card, and a field for any other information which will be stored as free-form text:

$$
\begin{array}{|l}
\hline
\quad \textit{DETAIL} \\
\hline
\textit{ccl} : AMOUNT \\
\textit{pin} : seq\ \mathbb{N} \\
\textit{info} : seq\ CHAR \\
\hline
\textit{pin} \in \{x : seq\ \mathbb{N}\ |\ \#x = 4\ \wedge\ ran\ x \subseteq 1..9\} \\
\hline
\end{array}
$$

i.e. the information will be modelled as an ordered sequence of characters, and the PIN is a member of the set of all 4-digit sequences of decimal digits. Finally, the CCL is defined as an *AMOUNT* where

$AMOUNT \triangleq \mathbb{Z}$

i.e. any integer number which represents a number of whole new pence.

If the customer does not have an ATM card, then the PIN number will be set to a rogue value:

$$| \ rogue : PIN$$

The third facet, $BALANCE$, has two components of the same type: the Available Balance and the Actual Balance for the bank account:

―――― $BALANCE$ ――――――――――――――――――
$actual : AMOUNT$
$avail : AMOUNT$
―――――――――――――――――
$actual \geq avail$

Note that the available balance must never exceed the actual balance—it would be erroneous if it did so.

The final facet $TRANLOG$ is modelled as a sequence of transaction elements:

$$TRANLOG \triangleq seq\ TRAN$$

―――― $TRAN$ ―――――――――――――――
$trantype : TYPE$
$timestamp : TIME$
$data : seq\ CHAR$

where

$$TYPE \triangleq \{\text{open}, \text{close}, \text{crcsh}, \text{crchq}, \text{drcsh}, \text{drchq}, \text{trsrc}, \text{trdst},\\ \text{ordstm}\}$$

i.e. $TYPE$ is the set of all possible types of transaction identifiers. $TIME$ is a timestamp indicating when a transaction took place.

The four facets will be indexed by the customer's account number using four partial functions. Each account number must map to exactly one tuple in each of the four functions, thus the domains of each of the functions will be equal. The account numbers will be members of the set *ACNO*.

<small>The nature of the partial function enforces the fact that each account number is associated with a unique customer.</small>

The basic types of this specification are as follows:

[*NAME, ADDR, PHONE, CHAR, TIME, ACNO*]

The details of the definitions of these types will be decided later in the development of the system.

The following is the specification of the state space of the banking system:

―――― *Bank* ――――――――――――――――――――
$id : ACNO \twoheadrightarrow IDENT$
$det : ACNO \twoheadrightarrow DETAIL$
$bal : ACNO \twoheadrightarrow BALANCE$
$log : ACNO \twoheadrightarrow TRANLOG$
――――――――――――――――――――――
$dom\ id = dom\ det = dom\ bal = dom\ log$

<small>Note that "dom id = dom det = dom bal..." is equivalent to "dom id = dom det ∧ dom det = dom bal ∧ ...".</small>

The initial state of the bank is when the bank has no customers:

―――― *InitBank* ―――――――――――――――
*Bank*
――――――――――――――
$id = \varnothing \wedge det = \varnothing \wedge bal = \varnothing \wedge log = \varnothing$

A global variable called *time* will hold the current time and date timestamp to the nearest one-hundredth of a second:

| $time : TIME$

The operations which alter the bank database can be classified into three types:

■ those which change the state of each of the four bank functions,
■ those which change the state of *id* and/or *det* (which are administration-type changes), and

- those which change the state of *bal* and *log* (which are bank transactions).

In order to make explicit the type of operation being performed, the following definitions will hold:

$BankAdmin \triangleq [\Delta Bank \mid bal' = bal \wedge log' = log]$

$BankTransaction \triangleq [\Delta Bank \mid id' = id \wedge det' = det]$

The first operation we will specify will be to introduce a new customer to the bank and open an account for this person. Each customer is assigned a unique account number for each new account. The customer opens an account with an initial (positive) amount of money, called the opening balance, which automatically becomes available to the customer. If the customer pays by cheque, then the cheque number will be entered as data. Any miscellaneous information is also entered into the customer's details:

```
___ NewCustomer _____
ΔBank
opening? : AMOUNT
iden? : IDENT
misc? : seq CHAR
data? : seq CHAR
an! : ACNO
_____
iden? ∉ ran id ∧ opening? > 0
(∃ n : ACNO; d : DETAIL; b : BALANCE; t : TRAN | n ∉ dom id • an! = n ∧
    d.ccl = 0 ∧ d.pin = rogue ∧ d.info = misc? ∧
    b.actual = b.avail = opening? ∧
    t.trantype = open ∧ t.timestamp = time ∧ t.data = data? ∧
    id' = id ∪ {n ↦ iden?} ∧
    det' = det ∪ {n ↦ d} ∧
    bal' = bal ∪ {n ↦ b} ∧
    log' = log ∪ {n ↦ ⟨t⟩})
```

In the predicate part, there is no conjunction symbol joining the two normal lines. Some people choose to omit the conjunction in Z schemas, the convention being that no symbol implies conjunction. Note that the same does not hold for disjunction—a disjunction symbol must always be shown when applicable. Conjunction has been shown in quantifications.

Sometimes an existing customer may wish to open a new separate account. In this situation, if the old account is given then the customer's identity will already be known:

```
┌─── OpenAnotherAccount ──────────────────────────────
│ ΔBank
│ opening? : AMOUNT
│ ac?, an! : ACNO
│ misc? : seq CHAR
│ data? : seq CHAR
│ ─────────────────────────────────────────────────
│ ac? ∈ dom id ∧ opening? > 0
│ (∃ n : ACNO; d : DETAIL; b : BALANCE; t : TRAN | n ∉ dom id • an! = n ∧
│     d.ccl = 0 ∧ d.pin = rogue ∧ d.info = misc? ∧
│     b.actual = b.avail = opening? ∧
│     t.trantype = open ∧ t.timestamp = time ∧ t.data = data? ∧
│     id' = id ∪ {n ↦ id(ac?)} ∧ det' = det ∪ {n ↦ d} ∧
│     bal' = bal ∪ {n ↦ b} ∧ log' = log ∪ {n ↦ ⟨t⟩})
└─────────────────────────────────────────────────────
```

There are two ways in which a customer can make a deposit, by cash or by cheque. Paying by cheque does not increase the Available Balance until the cheque is cleared:

```
┌─── Deposit ─────────────────────────────────────────
│ BankTransaction
│ ac? : ACNO
│ type? : {chq, cash}
│ sum? : AMOUNT
│ data? : seq CHAR
│ ─────────────────────────────────────────────────
│ ac? ∈ dom id ∧ sum? > 0
│ (∃ b1, b1 : BALANCE; t1, t2 : TRAN • t1.timestamp = t2.timestamp = time ∧
│     b1.actual = b2.actual = bal(ac?).actual + sum? ∧
│     b1.avail = bal(ac?).avail + sum? ∧
│     b2.avail = bal(ac?).avail ∧
│     t1.trantype = crcsh ∧ t2.trantype = crchq ∧
│     t1.data = t2.data = data? ∧
│ (type? = cash ⇒ bal' = bal ⊕ {ac? ↦ b1} ∧ log' = log ⊕ {ac? ↦ log(ac?)⌢⟨t1⟩})
│ (type? = chq ⇒ bal' = bal ⊕ {ac? ↦ b2} ∧ log' = log ⊕ {ac? ↦ log(ac?)⌢⟨t2⟩}))
└─────────────────────────────────────────────────────
```

Money can be withdrawn by cash or cheque only if there are sufficient funds in the Available Balance, or the excess does not exceed the CCL. If so, then the amount withdrawn will be subtracted from both balances:

---
__Withdrawal__
---
$BankTransaction$
$ac? : ACNO$
$type? : \{\texttt{chq}, \texttt{cash}\}$
$sum? : AMOUNT$
$data? : \text{seq } CHAR$

---
$ac? \in \text{dom } id \land sum? \leq bal(ac?).avail + det(ac?).ccl$
$(\exists\, b : BALANCE; t1, t2 : TRAN \bullet t1.timestamp = t2.timestamp = time \land$
$\quad b.actual = bal(ac?).actual - sum? \land$
$\quad b.avail = bal(ac?).avail - sum? \land$
$\quad t1.trantype = \texttt{drcsh} \land t2.trantype = \texttt{drchq} \land$
$\quad t1.data = t2.data = data? \land$
$\quad bal' = bal \oplus \{ac? \mapsto b\} \land$
$\quad (type? = \texttt{cash} \Rightarrow log' = log \oplus \{ac? \mapsto log(ac?)\frown\langle t1\rangle\})$
$\quad (type? = \texttt{chq} \Rightarrow log' = log \oplus \{ac? \mapsto log(ac?)\frown\langle t2\rangle\}))$
---

Ordering a statement is simply a matter of recording the transaction in the log. Every night, a batch run on the computer checks through the log for the previous 24-hour period, and carry out the necessary procedures for sending the statement to the customer. As there is no need to record any additional information about the transaction, a "null" value will be put in the log data field, defined globally as:

---
$null : \text{seq } CHAR$
---
$null = \langle\rangle$
---

---
__OrderStatement__
---
$BankTransaction$
$ac? : ACNO$

---
$ac? \in \text{dom } id$
$(\exists\, t : TRAN \bullet t.timestamp = time \land t.data = null \land t.trantype = \texttt{ordstm} \land$
$\quad log' = log \oplus \{ac? \mapsto log(ac?)\frown\langle t\rangle\} \land bal' = bal)$
---

Transfering money between accounts needs a source and destination account to be specified. At this level of the specification we will model this data as a sequence of characters. The same rules for withdrawing money from the source account apply as in withdrawing money in general:

─── *Transfer* ───────────────────────────────
*BankTransaction*
$src?, dst? : ACNO$
$sum? : AMOUNT$
$data? : \text{seq } CHAR$
────────────────────────────────
$\{dst?, src?\} \subseteq \text{dom } id \land src? \neq dst? \land sum? \leq bal(src?).avail + det(src?).ccl$
$(\exists\, b1, b1 : BALANCE;\ t1, t2 : TRAN \bullet t1.timestamp = t2.timestamp = time\ \land$
    $b1.actual = bal(dst?).actual + sum?\ \land$
    $b1.avail = bal(dst?).avail + sum?\ \land$
    $b2.actual = bal(src?).actual - sum?\ \land$
    $b2.avail = bal(src?).avail - sum?\ \land$
    $t1.trantype = \texttt{trdst} \land t2.trantype = \texttt{trsrc}\ \land$
    $t1.data = t2.data = data?\ \land$
$bal' = bal \oplus \{src? \mapsto b2, dst? \mapsto b1\}\ \land$
$log' = log \oplus \{src? \mapsto log(src?)\frown \langle t2 \rangle, dst? \mapsto log(dst?)\frown \langle t1 \rangle\})$
────────────────────────────────

Printing a statement of balance simply involves referencing the Actual Balance and Available Balance for the customer, and hence does not affect the state of the bank:

─── *BalanceEnquiry* ───────────────────────
$\Xi Bank$
$ac? : ACNO$
$balances! : BALANCE$
────────────────────────────────
$ac? \in \text{dom } id$
$balances! = bal(ac?)$
────────────────────────────────

Closing an account is a two stage process. The first stage involves registering the transaction, paying the customer an amount equal to their Available Balance, and informing them of the final balance when all cheques have cleared:

```
┌─ CloseStageOne ──────────────────────────────────
│ BankTransaction
│ ac? : ACNO
│ paidnow!, due! : AMOUNT
│ ─────────────────────────────────────────────────
│ ac? ∈ dom id ∧ bal(ac?).avail ≥ 0
│ (∃ b : BALANCE; t : TRAN • t.timestamp = time ∧ t.data = null ∧
│       b.actual = bal(ac?).actual − bal(ac?).avail ∧
│       b.avail = 0 ∧
│       t1.trantype = close ∧
│ log' = log ⊕ {ac?↦ log(ac?)⌢⟨t⟩} ∧ bal' = bal ⊕ {ac?↦ b})
│ paidnow! = bal(ac?).avail
│ due! = bal(ac?).actual − bal(ac?).avail
└──────────────────────────────────────────────────
```

Note that if the Available Balance is negative, then the customer will have to pay the bank before the account can be closed.

The second stage involves sending the customer any remaining money through the post (the difference between the Actual Balance and Available Balance when the account was closed in stage one) and a final notice of closure. This can only be carried out if all the customer's cheques have cleared, i.e. when the Available Balance and Actual Balance are the same. The account is then deleted from the main computer database:

```
┌─ CloseStageTwo ──────────────────────────────────
│ ΔBank
│ ac? : ACNO
│ due! : AMOUNT
│ addressee! : IDENT
│ ─────────────────────────────────────────────────
│ ac? ∈ dom id ∧ bal(ac?).avail = bal(ac?).actual
│ due! = bal(ac?).avail
│ addressee! = id(ac?)
│ id' = id ⩤ ac?
│ det' = det ⩤ ac?
│ bal' = bal ⩤ ac?
│ log' = log ⩤ ac?
└──────────────────────────────────────────────────
```

This is as far as we will specify the Bank operations in this case study. The following is a specification of the ATM operations:

# Z Specification of an ATM (extract)

The ATMs wait for a customer to insert an ATM card. When entered, the magnetic strip on the card provides the account number, and the customer is prompted for a valid 4-digit PIN number.

The ATM operations are essentially the same as their counterparts just specified for the human operated bank terminals. The following schema asserts the condition that a correct PIN number be entered before an ATM operation can be carried out:

```
┌─ ATMOp ─────────────────────────────
│ Bank
│ ac? : ACNO
│ pn? : seq N
├─────────────────────────────────────
│ pn? = det(ac?).pin
└─────────────────────────────────────
```

The customer can choose to withdraw up to £100, in £10 denominations, from the bank account if sufficient funds are available. The customer can only withdraw cash, and thus the type of transaction is always drcsh. The following schema asserts that an ATM withdrawal operation should be a valid ATM operation with a request for money in multiples of ten:

```
┌─ ATMDR ─────────────────────────────
│ ATMOp
│ sum? : AMOUNT
│ type? : {cheque, cash}
├─────────────────────────────────────
│ 1 ≤ (sum? div 10) ≤ 10
│ type? = cash
└─────────────────────────────────────
```

> The "div" operator is built-in to the Z Notation. It is an arithmetic operation such that "x div y" equals the rounded-down result of dividing x by y i.e. if x+y = z remainder r, then x div y = z (which is an integer).

The above schema can now be joined with the previously defined *Withdrawal* operation in order to completely specify an ATM withdrawal operation:

$$ATMWithdrawal \,\hat{=}\, ATMDR \land Withdrawal$$

In a similar way, the ATM request for a statement can be modelled as a valid ATM operation which also carries out the same procedure of ordering a statement as defined previously:

$$ATMOrderStatement \,\hat{=}\, ATMOp \land OrderStatement$$

An ATM balance enquiry can be modelled as a valid ATM operation and a request for a balance to be printed:

$$ATMBalanceEnquiry \,\hat{=}\, ATMOp \land BalanceEnquiry$$

## Z Specification of an the Banking Error Conditions

We will not cover the definitions of the error operations here. However, the same principles for specifying errors can be used as in Chapter Four. For example:

―― $InvalidACNO$ ――――――――――――――――――
$\Xi Bank$
$ac? : ACNO$
$report! : seq\ CHAR$
――――――
$ac? \notin dom\ id$
$report! = $ "Invalid Account Number"
―――――――――――――――――――――――――

―― $InsufficientFunds$ ――――――――――――――――
$\Xi Bank$
$ac? : ACNO$
$sum? : AMOUNT$
$report! : seq\ CHAR$
――――――
$sum? > bal(ac?).avail + det(ac?).ccl$
$report! = $ "Insufficient Funds"
―――――――――――――――――――――――――

┌─── *Correct* ──────────────────────────────
│ *report!* : *seq CHAR*
│ ──────────────────────────
│ *report!* = "Transaction Completed Successfully"
└────────────────────────────────────────────

Then, the new definition of *Withdrawal* would be as follows:

$FullWithdrawal \triangleq (Withdrawal \land Correct) \lor InsufficientFunds \lor InvalidACNO$

# APPENDIX B

# Answers To Exercises

## Exercise 2.1

The empty set contains no members, therefore its cardinality is zero.

## Exercise 2.2

a), e).

*Notes*

- b) is false because $B$ only has four member (one of which is a set).
- c) is false because 10 *is* a member of $B$.
- d) is false because the *singleton set* {4} is not a member of $A$.

## Exercise 2.3

a), b), d), e).

*Note*

- c is false because mauve is not a member of *rainbow*.

## Exercise 3.1

a)

| $P$ | $\neg P$ | $\neg(\neg P)$ |
|---|---|---|
| T | F | T |
| F | T | F |

b)

| $P$ | $\neg P$ | $P \vee (\neg P)$ |
|---|---|---|
| T | F | T |
| F | T | T |

c)

| $P$ | $\neg P$ | $P \wedge (\neg P)$ |
|---|---|---|
| T | F | F |
| F | T | F |

d)

| $P$ | $Q$ | $P \wedge Q$ | $\neg P$ | $\neg Q$ | $(\neg P) \vee (\neg Q)$ | $\neg(P \wedge Q)$ |
|---|---|---|---|---|---|---|
| T | T | T | F | F | F | F |
| T | F | F | F | T | T | T |
| F | T | F | T | F | T | T |
| F | F | F | T | T | T | T |

196

e)

| P | Q | $P \vee Q$ | $\neg P$ | $\neg Q$ | $(\neg P) \wedge (\neg Q)$ | $\neg(P \vee Q)$ |
|---|---|---|---|---|---|---|
| T | T | T | F | F | F | F |
| T | F | T | F | T | F | F |
| F | T | T | T | F | F | F |
| F | F | F | T | T | T | T |

## Exercise 4.1

```
┌─── InitReservoir ──────────────────────────────
│ Reservoir
│ ──────────────
│ level < 1800 ∧ max = 3000 ∧ min = 70
```

## Exercise 5.1

*IsAttendedBy* can be defined as *Attends*~, and has type

$CRSCDE \leftrightarrow STUDENT$.

## Exercise 5.2

For each lecturer $x$, the set of students who attend $x$'s lecture is given by

$\{y : STUDENT \mid y \ (Attends \ \S \ IsTaughtBy) \ x\}$.

## Exercise 5.3

For a relation $R : A \leftrightarrow B$ and a set $S : \mathbf{P}A$

$R^* = R \cup R^+ \cup \{x, y : A \mid x = y\}$

## Exercise 5.4

For a relation $R : A \leftrightarrow B$ and a set $S : \mathbf{P}B$

$R \triangleright S = \{x : A; y : B \mid x \ R \ y \wedge y \notin S\}$

## Exercise 5.5

For a relation $R : A \leftrightarrow B$, a set $S : \mathbf{P}A$, and a set $T : \mathbf{P}B$

$$S \triangleleft R = (A \setminus S) \triangleleft\!\!\!- R$$

$$R \triangleright T = R \triangleright\!\!\!- (B \setminus T)$$

# APPENDIX C

# Glossary

The following pages provide a convenient summary of the symbols used in this book along with cross-references to the pages the symbols were defined.

## SET THEORY

| NOTATION | DESCRIPTION | REF. |
|---|---|---|
| $x \in A$ | Set Membership. $x$ is a member of $A$. | p17 |
| $x \notin A$ | Nonmembership. | p17 |
| $A = B$ | Set Equality. | p18 |
| $A \subseteq B$ | Subset Relation. | p22 |
| $A \subset B$ | Proper Subset Relation. $A \subseteq B \wedge A \neq B$. | p22 |
| $\{\}$ | The Empty Set. | p20 |
| $\emptyset$ | The Empty Set. | p20 |

| NOTATION | DESCRIPTION | REF. |
|---|---|---|
| $x..y$ | Number Range | p24 |
| $A \times B$ | Cartesian Product | p33 |
| $\mathbf{P}A$ | Powerset. The set of all subsets of $A$. | p37 |
| $\{a, b, c\}$ | Set Display. | p15 |
| $\{x : T \mid P\}$ | Set Comprehension. | p25 |
| $\{x : T \mid P \bullet f\}$ | Set Comprehension by Form. | p35 |
| $A \cap B$ | Set Intersection. | p29 |
| $A \cup B$ | Set Union. | p28 |
| $A / B$ | Set Difference. | p29 |
| $\#A$ | Cardinality. The number of members of $A$. | p19 |

# LOGIC

| NOTATION | DESCRIPTION | REF. |
|---|---|---|
| $\neg P$ | Negation. The predicate which is true only when $P$ is false i.e."*not P*". | p48 |
| $P \wedge Q$ | Conjunction. The predicate which is true only when both $P$ and $Q$ are true i.e."*P and Q*". | p51 |
| $P \vee Q$ | Disjunction. The predicate which is false only when both $P$ and $Q$ are false i.e."*P or Q*". | p48 |
| $P \Rightarrow Q$ | Implication. The predicate which is true only when $P$ is false or both $P$ and $Q$ are true i.e."*if P then Q*". | p52 |
| $P \Leftrightarrow Q$ | Equivalence. The predicate which is true only when $P$ and $Q$ are both true or both false. | p46 |
| $(\forall D \mid Q \bullet P)$ | Universal Quantification. | p60 |

| $(\forall D \bullet P)$ | $(\forall D \mid true \bullet P)$ | p60 |
| $(\exists D \mid Q \bullet P)$ | Existential Quantification. | p58 |
| $(\exists D \bullet P)$ | $(\exists D \mid true \bullet P)$ | p57 |
| $(\exists_1 D \mid Q \bullet P)$ | Unique Quantification. | p59 |
| $(\exists_1 D \bullet P)$ | $(\exists_1 D \mid true \bullet P)$ | p59 |
| *true* | The predicate which is always true. | p45 |
| *false* | The predicate which is always false. | p45 |

## RELATIONS

| NOTATION | DESCRIPTION | REF. |
|---|---|---|
| $A \leftrightarrow B$ | Set of all binary relations between $A$ and $B$. | p107 |
| $x \mapsto y$ | Maplet. | p106 |
| $dom\ R$ | Domain of a relation. | p110 |
| $ran\ R$ | Range of a relation. | p111 |
| $id\ R$ | Identity Relation. | p112 |
| $R \mathbin{;} S$ | Relational Composition. | p116 |
| $R \circ S$ | Backward Relational Composition i.e. $R \circ S = S \mathbin{;} R$ | p117 |
| $S \triangleleft R$ | Domain Restriction. | p122 |
| $S \mathbin{\triangleleft\!\!\!-} R$ | Domain Subtraction. | p124 |
| $R \triangleright S$ | Range Restriction. | p123 |
| $R \mathbin{-\!\!\!\triangleright} S$ | Range Subtraction. | p124 |
| $R\sim$ | Relational Inversion. | p113 |
| $R(\!(S)\!)$ | Relational Image. | p114 |
| $R^+$ | Transitive Closure. | p120 |
| $R^*$ | Reflexive-Transitive Closure. | p120 |

# FUNCTIONS

| NOTATION | DESCRIPTION | REF. |
|---|---|---|
| $A \nrightarrow B$ | Set of all partial functions from $A$ to $B$. | p131 |
| $A \rightarrow B$ | Set of all total functions from $A$ to $B$. | p132 |
| $A \nrightarrowtail B$ | Set of all partial injections from $A$ to $B$. | p136 |
| $A \rightarrowtail B$ | Set of all total injections from $A$ to $B$. | p136 |
| $A \twoheadrightarrow B$ | Set of all partial surjections from $A$ to $B$. | p137 |
| $A \twoheadrightarrow B$ | Set of all total surjections from $A$ to $B$. | p137 |
| $A \rightarrowtail\!\!\!\twoheadrightarrow B$ | Set of all bijections from $A$ to $B$. | p137 |
| $f \oplus g$ | Function Override. | p133 |

# SEQUENCES

| NOTATION | DESCRIPTION | REF. |
|---|---|---|
| $seq\, A$ | Set of all sequences on $A$. | p142 |
| $iseq\, A$ | Set of all injective sequences on $A$. | p144 |
| $seq_1\, A$ | Set of all non-empty sequences on $A$. | p144 |
| $\langle a, b, c \rangle$ | Sequence Display. | p142 |
| $s \frown t$ | Concatenation. | p146 |
| $head\, s$ | Decomposition: the first element of $s$. | p147 |
| $tail\, s$ | Decomposition: a sequence of all but the first element in $s$. | p147 |
| $front\, s$ | Decomposition: a sequence of all but the last element in $s$. | p148 |
| $last\, s$ | Decomposition: the last element in $s$. | p147 |
| $rev\, s$ | Reverse of sequence $s$. | p149 |

| | | |
|---|---|---|
| $s \upharpoonright A$ | Filtering. | p150 |
| $\frown\!\!\!/ s$ | Distributed concatenation. | p151 |
| *disjoint s* | Disjointness. | p152 |
| *s partition A* | Partitions. | p153 |

# BAGS

| NOTATION | DESCRIPTION | REF. |
|---|---|---|
| *bag A* | Set of all bags on $A$. | p157 |
| $[a, b, c]$ | Bag Display. | p156 |
| *x in b* | Bag Membership. Test if $x$ is in $b$. | p157 |
| *count b x* | Bag Count. Number of $x$'s in $b$. | p158 |
| *items s* | Bag of a sequence $s$. | p158 |
| $b \uplus c$ | Bag Union. | p159 |

# ARITHMETIC

| NOTATION | DESCRIPTION | REF. |
|---|---|---|
| **Z, N** | Integers, Natural Numbers. | p23 |
| - | Negation. | - |
| +, -. | Addition, Subtraction. | - |
| *, *div*, *mod* | Multiplication, Integer Division, Modulo. e.g. when $b \neq 0$ then $a = (a\ div\ b) * b + a\ mod\ b$. | - |
| $<, \leq, >, \geq$ | Relational Comparision. | p44 |

# Index

## A
abbreviation definitions 172
abstract data types 166
abstraction 15, 26, 67
actual generic parameters 170
after state 78
alphabet 175
and (logical) 51
antecedent 52
arguments 130
axiomatic descriptions 91
axioms 69

## B
backwards relational composition 117
bags 155-160, 161
- bag membership 157
- bag union 159
basic types 93
before state 78
bijections 137

binding formation 164
bound variables 56

## C
cardinality 19
cartesian product 33, 160
changes in state 31
complement 47
communication 3
Computer Science 4
concatenation 146
conjunction 51
connectives 46
consequent 52
constraints 46, 91
counting objects 157, 173

## D
data types 160
databases 27
declarations 57, 67

decomposition 148
decorations 31
deductive apparatus 176
default constraint 72
deterministic behaviour 138
disjoint sets 152
disjunction 48
distributed concatenation 151
domain 110
domain restriction 122
domain subtraction 124, 133
dynamic properties 78

## E
elements 17
elementary predicates 44
equality 18
equivalence 16, 46
equivalence transformations 47
example specification 95
existential quantification 56

## F
*false* 45
filtering 46, 58, 150
finite sets 23
formal generic parameters 170
formal languages 175
formal specifications 4
formal systems 176
free types 174
functions 127-138, 161
– function application 129
– function overriding 133
further reading 11

## G
generalised union 174
generic constants 171
global variables 89

## H
hierarchical models 21
homogeneous relations 115
Human-Computer Interaction 2

## I
identifiers 16
implication 51
indexing 144
infinite sets 24
initial state 75
injections 136
injective sequences 144
integers 23
invariants 73

## J
joining sets 29

## L
lambda expressions 172
length 145
LHS 16
local variables 89
logic 43-62, 160
– laws of and-simplification 51
– laws of or-simplification 48

## M
maplets 106
models 72
model-based specifications 68

modifiers 46
modularity 70
modulo operator 26

## N
natural numbers 23
negation 47
non-empty sequences 143
number range 24

## O
objects 14, 66
operands 17, 53
operators 17
or (logical) 48
ordered pairs 32, 106, 160

## P
paragraphs 94
partial functions 130
partitions 153
postconditions 79
powersets 37
precedence rules 54
preconditions 79
predicates 44
proofs 43, 176
proper subsets 22

## R
range 56, 111
range restriction 122
range subtraction 124
reference manual for Z 12
reflexivity 115
reflexive-transitive closure 119
relations 105-125, 129, 161
– identity relation 112

– relation symbol 107
– relational composition 116
– relational image 114
– relational inversion 112
– relational iteration 118
relational operators 44, 109
relationships 68, 105
repetitions 20
RHS 16

## S
schemas 68-94
– generic schemas 170
– global signatures 90
– predicate part 68
– schema composition 174
– schema expressions 85
– schema operators 85
– schema reference 76
– schema types 165
– signature 68
scope 89
selection 167
semantic rules 175
semantics 172
sequences 141-154
set theory 13-41, 160
– empty set 20
– set comprehension 25
– set comprehension by form 35
– set difference 29
– set display 15
– set intersection 29
– set membership 17, 28
– set union 28
– singleton sets 20, 30
Software Engineering 2

*Specification Case Studies* 12
state space 73
static properties 73
strong (predicates) 92
subsets 22
summation quantifier 56
surjections 137
symmetry 115
syntax rules 175

# T
tautologies 50
total functions 132
transitivity 115
transitive closure 119
*true* 45
truth tables 49
tuples 32
types 25, 37, 57, 68

# U
undecidability 61
unique existence 59
universal quantification 59
unordered objects 19

# V
variables 66
Venn diagrams 38

# W
weaker (predicates) 92
well-formed formulae 175

# Z
Z 8
– Z Language 175
– Z specification document 94

## GENERAL COMPUTING BOOKS
Compiler Physiology for Beginners, M Farmer, 279pp, ISBN 0-86238-064-2
Concise Dictionary of Computing and Information Technology, D Lynch, 380 pages, ISBN 0-86238-268-8
File Structure and Design, M Cunningham, 211pp, ISBN 0-86238-065-0
Information Technology Dictionary of Acronyms and Abbreviations, D Lynch, 270pp, ISBN 0-86238-153-3
Project Skills Handbook, S Rogerson, 143pp, ISBN 0-86238-146-0

## PROGRAMMING LANGUAGES
An Intro to LISP, P Smith, 130pp, ISBN 0-86238-187-8
An Intro to OCCAM 2 Programming: 2nd Ed, Bowler, et al, 109pp, ISBN 0-86238-227-0
BASIC Applications Programming, Parker, 312pp, ISBN 0-86238-263-7
BASICALLY MODULA-2, Walmsley/Williams, 228pp, ISBN 0-86238-270-X
C Simply, M Parr, 168pp, ISBN 0-86238-262-9
Cobol for Mainframe and Micro: 2nd Ed, D Watson, 177pp, ISBN 0-86238-211-4
Comparative Languages: 2nd Ed, J R Malone, 125pp, ISBN 0-86238-123-1
Fortran 77 for Non-Scientists, P Adman, 109pp, ISBN 0-86238-074-X
Fortran 77 Solutions to Non-Scientific Problems, P Adman, 150pp, ISBN 0-86238-087-1
Fortran Lectures at Oxford, F Pettit, 135pp, ISBN 0-86238-122-3
LISP: From Foundations to Applications, G Doukidis et al, 228pp, ISBN 0-86238-191-6
Programming for Change in Pascal, D Robson, 272pp, ISBN 0-86238-250-5
Prolog versus You, A Johansson, et al, 308pp, ISBN 0-86238-174-6
Simula Begin, G M Birtwistle, et al, 391pp, ISBN 0-86238-009-X
Structured Programming with COBOL & JSP: Vol 1, J B Thompson, 372pp, ISBN 0-86238-154-1, Vol 2, 354pp, ISBN 0-86238-245-9
The Intensive C Course: 2nd Edition, M Farmer, 186pp, ISBN 0-86238-190-8
The Intensive Pascal Course: 2nd Edition, M Farmer, 125pp, ISBN 0-86238-219-X

## ASSEMBLY LANGUAGE PROGRAMMING
Coding the 68000, N Hellawell, 214pp, ISBN 0-86238-180-0
Computer Organisation and Assembly Language Programming, L Ohlsson & P Stenstrom, 128pp, ISBN 0-86238-129-0
What is machine code and what can you do with it? N Hellawell, 104pp, ISBN 0-86238-132-0

## PROGRAMMING TECHNIQUES
An Introduction to Z, Imperato, 208pp, ISBN 0-86238-289-0
Discrete-events simulations models in PASCAL/MT+ on a microcomputer, L P Jennergren, 135pp, ISBN 0-86238-053-7
Information and Coding, J A Llewellyn, 152pp, ISBN 0-86238-099-5
JSP - A Practical Method of Program Design, L Ingevaldsson, 204pp, ISBN 0-86238-107-X
Modular Software Design, M Stannett, 136pp, ISBN 0-86238-266-1
Simulation Modelling, Paul/Balmer, 154pp, ISBN 0-86238-280-7
Software Engineering Fundamentals, Ingevaldsson, 248pp, ISBN 0-86238-103-7

**Programming for Beginners: the structured way,** D Bell & P Scott, 178pp,
ISBN 0-86238-130-4
**Software Engineering for Students,** M Coleman & S Pratt, 195pp,
ISBN 0-86238-115-0
**Software Taming with Dimensional Design,** M Coleman & S Pratt, 164pp,
ISBN 0-86238-142-8

## MATHEMATICS AND COMPUTING
**Fourier Transforms in Action,** F Pettit, 133pp, ISBN 0-86238-088-X
**Generalised Coordinates,** L G Chambers, 90pp, ISBN 0-86238-079-0
**Hyperbolic Problems Vols 1&2,** Engquist/Gustafson, ISBN 0-86238-285-8
**Linear Programming: A Computational Approach: 2nd Ed,** K K Lau, 150pp,
ISBN 0-86238-182-7
**Numerical Methods of Linear Algebra,** S Laflin, 170pp, ISBN 0-86238-151-7
**Statistics and Operations Research,** I P Schagen, 300pp, ISBN 0-86238-077-4
**Teaching of Modern Engineering Mathematics,** L Rade (ed), 225pp,
ISBN 0-86238-173-8
**Teaching of Statistics in the Computer Age,** L Rade (ed), 248pp, ISBN 0-86238-090-1
**The Essentials of Numerical Computation,** M Bartholomew-Biggs, 241pp,
ISBN 0-86238-029-4

## DATABASES AND MODELLING
**Computer Systems Modelling & Development,** D Cornwell, 291pp,
ISBN 0-86238-220-3
**An Introduction to Data Structures,** B Boffey, D Yates, 250pp, ISBN 0-86238-076-6
**Database Analysis and Design: 2nd Ed,** H Robinson, 378pp, ISBN 0-86238-018-9
**Databases and Database Systems: 2nd Ed,** E Oxborrow, 256pp, ISBN 0-86238-091-X
**Data Bases and Data Models,** B Sundgren, 134pp, ISBN 0-86238-031-6
**Text Retrieval and Document Databases,** J Ashford & P Willett, 125pp,
ISBN 0-86238-204-1
**Information Modelling,** J Bubenko (ed), 687pp, ISBN 0-86238-006-5

## UNIX
**An Intro to the Unix Operating System: 2 Ed,** C Duffy, 152pp, ISBN 0-86238-271-8
**Operating Systems through Unix,** G Emery, 96pp, ISBN 0-86238-086-3

## SYSTEMS ANALYSIS & SYSTEMS DESIGN
**Systems Analysis and Development: 3rd Ed,** P Layzell & P Loucopoulos, 284pp,
ISBN 0-86238-215-7
**SSADM Techniques: Version 4,** Lejk, et al, 350pp, ISBN 0-86238-224-6
**Computer Systems: Where Hardware meets Software,** C Machin, 200pp,
ISBN 0-86238-075-8
**Microcomputer Systems: hardware and software,** J Tierney, 168pp,
ISBN 0-86238-218-1
**Distributed Applications and Online Dialogues: a design method for application systems,** A Rasmussen, 271pp, ISBN 0-86238-105-3